Hockey and AI

The Future of the Game

By
Oakley Hunter

Hockey and AI

The Future of the Game

Table of Contents

Introduction

Field hockey has always been a sport defined by its rich history, tactical complexity, and the physical prowess of its players. However, in recent years, the game has undergone a quiet transformation, fueled by rapid advances in artificial intelligence (AI). Today, AI is not just an abstract concept confined to laboratories and computer science textbooks; it has found its way into the field, reshaping the very fabric of how the sport is played, coached, and analyzed.

As AI-powered tools become increasingly mainstream, their effects on field hockey are profound and multifaceted. This book aims to peel back the layers of this transformation, offering an in-depth look at how AI is making its mark on the field. Whether you're a player keen to optimize your skills, a coach looking for an edge in strategy, or an enthusiast eager to understand the technological underpinnings of your favorite sport, this book is designed for you.

To start, it's essential to understand why AI is particularly suited to revolutionize field hockey. Unlike traditional methods that may rely on intuition or static data, AI leverages dynamic algorithms to offer real-time insights and predictions. This capability aligns perfectly with the fast-paced, ever-changing nature of field hockey, where split-second decisions often determine the outcome of a match.

The enormous volume of data generated during games—such as player movements, ball trajectories, and team formations—provides fertile ground for AI. These large datasets are the raw materials that AI systems need to learn, adapt, and offer actionable insights. By analyzing patterns and trends that are often invisible to the human eye, AI provides a new lens through which the game can be understood and improved.

This book starts with a historical overview of field hockey, mapping its evolution from ancient times to the present day, and then dives into the basics of artificial intelligence, laying a foundation for readers unfamiliar with the technology. From there, it dissects various applications of AI in field hockey. These range from personalized training programs and real-time game strategy analytics to injury prevention and fan engagement.

One of the most exciting areas is AI-driven training. Imagine a training regimen so precisely tailored to an individual player's strengths and weaknesses that it's like having a personal coach available 24/7. AI can analyze past performances, identify areas needing improvement, and suggest drills to enhance specific skills. This personalized approach can drastically reduce the time needed to achieve mastery.

In addition to training, AI is changing the strategic aspects of the game. Coaches can now access sophisticated predictive models that simulate various game scenarios, helping them make data-driven decisions. Whether it's choosing the optimal lineup, devising new plays, or making in-game adjustments, AI adds a layer of precision and foresight that was previously unimaginable.

Performance analysis has also taken a quantum leap. With AI, it's possible to track a vast array of performance metrics over time, transforming how players and coaches interpret data. Video analysis powered by AI not only offers detailed breakdowns of matches but also generates actionable insights, helping teams refine their strategies and tactics continuously.

Injury prevention is another critical aspect where AI's impact is being keenly felt. Wearable technology equipped with AI can monitor physiological data, predicting potential injuries before they occur. By alerting players and coaches to the early signs of fatigue or stress, these systems can prolong careers and maintain player health.

Recruiting and talent identification have also seen significant advancements. AI's ability to assess large datasets quickly and accurately means that scouts can now evaluate more players in less time. This leads to more informed decisions and the discovery of hidden talents who might otherwise have been overlooked.

The fan experience is perhaps one of the most visible areas where AI is making waves. From AI-enhanced commentary during broadcasts to virtual and augmented reality experiences, fans are now more engaged and connected to the sport. These technologies not only make watching games more enjoyable but also provide deeper insights into the game's nuances.

Despite its numerous benefits, the integration of AI into field hockey isn't without challenges. Ethical considerations such as data privacy, fairness, and the balance between human and machine judgment are critical issues that need careful deliberation. This book will address these concerns, offering a balanced perspective on the ethical landscape of AI in sports.

Finally, we'll present real-world case studies that illustrate the successful implementation of AI in field hockey and other sports. These examples will provide practical insights and lessons learned, making it easier for teams and organizations to adopt AI technologies.

The concluding chapters will offer guidance on how to integrate AI into field hockey programs effectively. From initial steps to train coaches and players to evaluating the success of these initiatives, you'll find actionable strategies to ensure smooth implementation.

As we look towards the future, it's clear that AI will continue to evolve, bringing with it new innovations that will further transform field hockey. By staying informed and embracing these technologies, players, coaches, and fans can be part of this exciting journey, shaping the future of the sport.

Chapter 1:
The Evolution of Field Hockey

Field hockey, a sport with ancient roots dating back to civilizations such as Egypt and Persia, has undergone a remarkable evolution to become the fast-paced, dynamic game we recognize today. From rudimentary wooden sticks and natural turfs, the transition to advanced composite materials and synthetic pitches has significantly enhanced performance and spectator experience. The introduction of technology has revolutionized training methods, with wearable sensors and video analysis offering deeper insights into player performance. Equally transformative, the development of synthetic pitches has not only altered gameplay speed but also persistence, making the sport more accessible around the world. These advancements showcase how field hockey's rich history continues to merge with modern innovations, setting the stage for even greater leaps forward in the era of artificial intelligence.

Historical Context

Field hockey's rich and varied history provides an intriguing backdrop for understanding its evolution. Originating over 4,000 years ago in Ancient Egypt, the game was far removed from the high-speed, tech-enhanced version played today. The early versions of field hockey were rudimentary and played with crude, curved sticks and balls made from leather or tightly wound wool. These games, often depicted in ancient artwork, show a sport that was less about organized competition and more about communal joy and physical exercise.

The modern game of field hockey has roots in 19th-century England. It was during this time that the sport began to develop formal rules and standardizations. In the mid-1800s, English schools like Eton and Harrow adopted field hockey, which helped it gain popularity among young athletes. By 1886, the foundations of the first governing body, the Hockey Association in London, were laid. This organization aimed to refine the rules, making the sport more competitive and accessible.

A significant historical development occurred when field hockey was introduced to the Olympics. The sport made its debut at the 1908 London Games, although it wasn't included in the subsequent 1912 Stockholm Games. It returned in 1920 and has been a staple ever since. The inclusion in the Olympics not only increased the sport's visibility but also emphasized the need for formalized training and strategic advancements.

In the early 20th century, field hockey spread from England to its colonies including India, Australia, and various African nations. The Indian subcontinent, in particular, embraced the sport with passion and skill. India's hockey team dominated the Olympics from 1928 to 1956, winning six consecutive gold medals. These victories were not just about athletic excellence; they came to symbolize national pride and identity. The prowess of these players inspired generations, and many of the techniques and strategies developed during this period remain fundamental to today's game.

Women's field hockey also has a storied history. The first recorded women's field hockey match took place in 1895 in England, and the sport grew rapidly in the early 20th century. The formation of the All England Women's Hockey Association in 1895 was pivotal in promoting the sport among women. However, it wasn't until 1980 that women's field hockey was included in the Olympic Games, offering female athletes the same international platform as their male counterparts to showcase their talents.

As we moved into the latter half of the 20th century, field hockey began to see significant advances in equipment and playing surfaces. The introduction of synthetic turf in the 1970s was revolutionary. This change from natural grass to artificial turf increased the game's speed and technical demands. The 1976 Montreal Olympics was the first time the sport was played on an artificial surface, fundamentally altering how the game was played. Players had to adapt to quicker ball speeds and different tactical plays, prompting a shift towards more skill-oriented gameplay.

Technology began making its mark on field hockey in various ways beyond just playing surfaces. Advances in stick materials, ranging from wood to composite materials like fiberglass and carbon fiber, revolutionized how players controlled the ball and executed shots and passes. Protective gear also improved, enhancing player safety without compromising performance.

By the time we entered the 21st century, the fusion of technology and tradition became more evident. The introduction of high-definition video analysis, GPS tracking systems, and performance metrics allowed coaches to refine tactics and strategies by analyzing players' movements and behaviors in unprecedented detail. Training methods evolved to include data-driven insights, significantly impacting player development and team performance.

As field hockey continues to evolve, understanding its historical context is crucial. The steps taken in the past lay the foundation for the current innovations in artificial intelligence and technology. Appreciating the journey from ancient Egypt's simple wooden sticks to today's AI-enhanced coaching tools provides a deeper perspective on how far the sport has come and where it is headed. Each historical milestone, each technological innovation, contributes to the rich tapestry of field hockey, making it an ever-evolving and dynamic sport.

Advances in Equipment and Training

Field hockey has come a long way since its early days, and much of this transformation can be accredited to remarkable advances in equipment and training. These developments have not only made the game faster and safer but also more inclusive and competitive. Both players and coaches benefit from these innovations, allowing for a higher level of skill and performance on the field.

At the heart of these advances is the evolution of equipment, which has seen significant changes over the decades. The hockey stick, once a simple wooden implement, has been radically transformed. Modern sticks are often made of composite materials like fiberglass, carbon fiber, and Kevlar, making them lighter, stronger, and more durable. This shift has allowed players to execute more complex maneuvers with greater precision and less strain on their bodies.

Apart from sticks, the hockey ball itself has undergone changes. Originally made of leather, today's balls are crafted from plastic composites designed to be both durable and safe. The consistency in ball performance reduces the risk of unpredictable bounces, contributing to a more stable and flowing game.

Protective gear has also seen substantial improvements. Goalkeepers now wear lightweight pads and helmets made from advanced materials that offer superior protection without compromising agility. Field players have access to protective gear, such as shin guards and mouthguards, that are both highly effective and comfortably designed.

The introduction of artificial turf has been another game-changer. Natural grass fields, vulnerable to weather conditions, have largely been replaced by synthetic turf. This provides a consistent playing surface that greatly enhances the speed and reliability of the game. Field maintenance becomes less labor-intensive, and the playing conditions remain optimal regardless of weather.

Training methods have evolved alongside equipment innovations. Today, coaches leverage technology to design training programs that are highly tailored to individual needs. High-speed cameras and motion capture technology allow for detailed analysis of a player's technique, identifying strengths and areas for improvement with pinpoint accuracy.

Fitness training is now a science. Wearable fitness trackers monitor heart rate, movement patterns, and physical exertion, providing real-time data that can inform everything from a player's in-game decisions to their recovery protocols. Strength and conditioning programs have become more sophisticated, using data-driven approaches to ensure athletes are at their physical peak.

The rise of video analysis cannot be overlooked. Teams routinely use video footage to break down matches, analyze opponent strategies, and refine their game plans. Independent software tools equipped with artificial intelligence capabilities can process videos to detect patterns, identify key moments, and even suggest tactical adjustments. This level of analysis, previously unheard of, offers a competitive edge by turning raw data into actionable insights.

Moreover, wearable technology has become indispensable in modern training regimens. Smart clothing with embedded sensors tracks a wide range of metrics such as muscle load and joint movement, helping to prevent injuries and optimize performance. These insights enable the development of targeted training programs that improve strength, flexibility, and endurance while minimizing the risk of injury.

Physical conditioning is only part of the equation. Mental preparation has gained prominence, with mindfulness and psychological resilience becoming key focus areas. Virtual reality (VR) and augmented reality (AR) simulations let players experience game scenarios in a controlled environment, enhancing their decision-making skills and mental toughness.

Nutrition also plays a crucial role in modern training. Nutritional plans are now highly personalized, often designed in collaboration with dietitians and using data from biometric analyses. Proper nutrition aids in faster recovery times, better endurance, and overall improved performance.

Training facilities themselves have become more advanced. High-tech gyms, equipped with the latest in fitness and rehabilitation technology, provide all-inclusive environments tailored to the needs of elite athletes. Additionally, advancements in biomechanics have led to the creation of customized training equipment that targets specific muscle groups and movements relevant to the sport.

Another revolutionary aspect is the use of drones in training sessions. These airborne devices capture extensive footage from various angles, offering a unique perspective not accessible through traditional means. Coaches can review this footage to provide players with unparalleled feedback about positioning and movement on the field.

Finally, the role of data analytics cannot be overstated. Training programs are now constructed based on complex algorithms that consider a wide range of factors, from past performance to individual fitness levels. Machine learning systems analyze performance data to continuously refine and enhance training protocols, ensuring they are as effective as possible.

In essence, the advances in equipment and training have redefined what is possible in field hockey. The integration of cutting-edge technology, data analysis, and improved materials has elevated the sport to new heights. These innovations provide players with the tools they need to excel and offer coaches deep insights that were previously unimaginable. This rapid evolution ensures that field hockey remains not only relevant but also increasingly compelling and competitive for years to come.

The Role of Technology in Modern Field Hockey

Field hockey, like many other sports, has experienced a significant transformation due to advancements in technology. The digitization and integration of new tools have revolutionized how the game is played, analyzed, and enjoyed. The introduction of these technologies has not only enhanced the skills and performance of players but also led to more strategic and efficient coaching methods.

One of the key ways technology has impacted field hockey is through the development of advanced equipment. Modern field hockey sticks, for example, are now made from composites of fiberglass, Kevlar, and carbon fiber. This blend of materials allows for a lighter stick that offers better ball control and improved hitting power. As a result, players can execute more precise and powerful plays that were previously difficult to achieve with traditional wooden sticks.

Another major area where technology has made an impact is in training and performance monitoring. Through wearable technology like GPS trackers and heart rate monitors, coaches and analysts can gather real-time data on a player's performance. This data provides valuable insights into an athlete's physical condition, enabling tailored training programs that address specific needs. For instance, a coach can adjust a player's regimen to focus on endurance if the data shows they tire quickly during matches.

Video analysis tools are also a game-changer in modern field hockey. High-definition cameras and specialized software allow for detailed analysis of games, both in real time and after the fact. Coaches and players can dissect every play, identifying strengths and weaknesses that might not be apparent during live action. Video analysis helps in creating more effective game strategies and improving individual player performance.

Additionally, artificial intelligence (AI) has begun to play a critical role in the sport. AI-driven tools can process vast amounts of data to uncover patterns and trends that humans might overlook. For example, machine learning algorithms can analyze past matches to predict future performance outcomes, offering strategic advantages to teams willing to embrace this technology.

Moreover, technology has democratized access to top-notch training regardless of geographical location. Online platforms and mobile applications offer drills, tutorials, and even virtual coaching sessions. Players who once had limited access to professional training can now learn and improve their skills from anywhere in the world, leveling the playing field in a way that was unimaginable a few decades ago.

Field conditions and player safety have also been enhanced through technological advancements. Turf manufacturers have developed advanced synthetic fields that offer consistent performance and reduce the risk of injuries. These modern surfaces are engineered to provide optimal ball roll and grip, mimicking the consistency of natural grass but without the drawbacks of wear and weather dependency.

The integration of automated systems and robotics in off-field activities also contributes to the sport's evolution. Automated ball feeders and rebound boards are commonly used in practice sessions, allowing players to practice shots and reflexes in ways that simulate real-game scenarios. Such technologies provide consistent and repetitive training, honing skills that are crucial during actual matches.

The blend of data analytics and wearable technology has made conditioning and rehabilitation more scientific and effective. By continuously monitoring player metrics, these technologies can predict injuries before they occur, allowing for preventative measures to be taken. Post-injury, data-driven rehabilitation programs ensure that players recover fully and return to the field in optimal condition.

Off the field, technology has created a more immersive and interactive experience for fans. High-definition broadcasts, instant replays, and augmented reality features enrich the viewing experience, making it more engaging and enjoyable. Social media platforms and live-streaming services provide multiple avenues for fans to follow their favorite teams and players, fostering a deeper connection with the sport.

However, with all these advancements, it's crucial to acknowledge that technology alone doesn't define the game. At its core, field hockey remains a sport that relies on human skill, hard work, and passion. Technology acts as a catalyst, enhancing these fundamental aspects rather than replacing them.

Looking ahead, the continued integration of technology promises to make field hockey more dynamic and inclusive. Emerging tech trends like virtual reality training, AI-driven strategy simulations, and even more advanced performance tracking wearables hold the potential to further elevate the game. The fast-paced evolution of these technologies ensures that field hockey will keep pushing the boundaries of what's possible, making it an exciting time to be part of the sport.

In conclusion, the role of technology in modern field hockey is multifaceted and far-reaching. From improving equipment and training methods to enhancing player safety and fan engagement, technological advancements have indelibly changed the face of the game. As we forge ahead, embracing these innovations will be key to unlocking new levels of performance and enjoying the sport in previously unimaginable ways.

Chapter 2:
Understanding Artificial Intelligence

Artificial Intelligence (AI) is reshaping the landscape of field hockey in ways previously unimaginable, serving as a transformative force that blends seamlessly with traditional techniques. At its core, AI involves the simulation of human intelligence processes by machines, often through learning, reasoning, and self-correction. With machine learning and deep learning as its driving engines, AI can analyze massive datasets, recognize patterns, and make informed decisions with incredible speed and accuracy. Across various sports, AI has already made remarkable strides, enhancing performance analysis, strategy formulations, and injury prevention. For athletes, coaches, and analysts in field hockey, understanding these foundational concepts is crucial to leveraging AI effectively. This chapter explores the essence of AI, delineating its primary components and setting the stage for its specific applications in field hockey. Through this exploration, readers will gain a clear picture of how AI's promise can be realized on the turf, fostering a new era of enhanced gameplay and strategic depth.

Basic Concepts of AI

Artificial Intelligence, or AI, is a term that's increasingly becoming ingrained in our daily lives, and its implications for sports, including field hockey, are profound. At its core, AI involves creating systems that can perform tasks that typically require human intelligence. These tasks range from simple actions like recognizing speech to more complex processes like making strategic decisions. Understanding AI's basic concepts is the first step to appreciating its transformative potential in field hockey.

One of the fundamental elements of AI is the concept of algorithms. Algorithms are sets of rules or instructions that a computer follows to complete a task. In the context of AI, these algorithms are designed to learn and adapt from data, which means they can improve their performance over time without explicit programming. For instance, an AI system might analyze thousands of hours of field hockey footage to identify patterns that even seasoned coaches might miss.

Another key concept to grasp is machine learning. This subset of AI involves teaching computers to learn from and make decisions based on data. There are various types of machine learning, including supervised learning, unsupervised learning, and reinforcement learning. Supervised learning involves training the AI with labeled data so it can make predictions or decisions. Unsupervised learning, on the other hand, deals with data that hasn't been labeled, and the system tries to identify patterns or groupings within it. Reinforcement learning uses a system of rewards and punishments to "teach" the AI to achieve a specific goal.

An even more advanced subset of machine learning is deep learning. This approach uses neural networks with many layers—hence the term "deep." These neural networks are designed to mimic the human brain's structure and function, allowing them to process vast amounts of data and make complex decisions. In sports, deep learning can be employed to analyze player movements, predict game outcomes, and even develop new training methods that were previously unimaginable.

AI isn't just about crunching numbers and analyzing data; it also involves aspects of perception and interaction. Natural Language Processing (NLP) is one such area, focusing on the interaction between computers and human language. This allows AI systems to understand, interpret, and generate human language in a way that feels natural. In field hockey, NLP can be used to analyze commentary, provide real-time translation, or enhance fan engagement through interactive voice-assistants.

Computer vision is another critical area of AI that has substantial implications for field hockey. Computer vision enables machines to interpret and make decisions based on visual information. Using complex algorithms, AI systems can analyze video feeds to track player movements, assess playing techniques, and even identify potential injuries. The insights gained from computer vision can lead to more effective training programs, improved game strategies, and enhanced player safety.

Understanding these basic concepts isn't just academic; it has practical applications that can revolutionize how field hockey is played, coached, and analyzed. For instance, AI-driven video analysis can provide real-time feedback during matches, highlighting tactical adjustments that coaches can implement immediately. This dynamic approach to coaching and gameplay can offer a competitive edge that traditional methods simply can't match.

Moreover, AI's ability to handle and analyze big data means it can consider a vast array of variables when making a decision. Whether it's predicting the likelihood of a successful penalty shot or optimizing team formations, AI can sift through massive datasets to arrive at the most effective solutions. This level of analysis was unthinkable just a few years ago but is now within reach thanks to advancements in AI technology.

It's also worth noting the role of AI in off-field activities. For example, AI can assist in scouting and talent identification by analyzing young players' performances and extracting metrics that indicate future potential. This capability can democratize the scouting process, making it easier to identify hidden gems who might otherwise go unnoticed.

But AI isn't just about cold calculations and data. One of its most exciting aspects is its potential to enhance the human element in field hockey. By taking over mundane tasks and providing deeper insights, AI frees up coaches and players to focus on what they do best. Imagine a scenario where a coach doesn't have to spend hours reviewing game footage because an AI system has already highlighted the key moments and provided actionable insights. This allows coaches to focus on developing strategies and building team cohesion, areas where human intuition and experience are irreplaceable.

Another exciting development is using AI to personalize training programs for individual players. Each player has unique strengths, weaknesses, and learning styles. AI can tailor training programs to suit these individual characteristics, thereby maximizing each player's potential. This personalized approach isn't just beneficial for professional athletes; it can be a game-changer for young, aspiring players working their way up.

The journey to harnessing AI's full potential in field hockey is just beginning, but the foundational concepts are already transforming the sport. From algorithms and machine learning to NLP and computer vision, AI offers tools that can enhance every aspect of the game. As these technologies continue to evolve, their impact will only become more profound, making the game faster, smarter, and more exciting for everyone involved.

Machine Learning and Deep Learning

Field hockey, like many sports, thrives on strategy, speed, and precision. One of the most transformative innovations in recent years has been the incorporation of artificial intelligence in the game, particularly through machine learning and deep learning technologies. For field hockey players, coaches, and analysts, the impact of these technologies is multifaceted and profound.

Machine learning serves as the backbone of AI in sports. It involves the use of algorithms that can learn from and make predictions based on data. Applied to field hockey, machine learning can analyze vast amounts of performance data, from player positioning to stick handling techniques. Contrast this with traditional methods of performance evaluation, which often relied on the subjective judgment of coaches and analysts.

With machine learning, data doesn't just sit there; it speaks. Patterns and trends become evident in ways never before possible. For example, by analyzing game footage over an entire season, machine learning algorithms can identify specific instances where a player's performance declined, linking these moments to particular conditions or opponent strategies. It's like having a personal analyst who's always vigilant and never misses a detail.

Deep learning, a subset of machine learning, adds another layer of sophistication. It uses neural networks with many layers (hence "deep") to model complex patterns in data. In the context of field hockey, deep learning can power more advanced applications like video analysis and predictive analytics. Imagine a scenario where coaches have access to a system that not only breaks down video footage of games but also provides actionable insights on player fatigue levels, missed opportunities, and likely future performance.

These deep learning systems are trained on thousands of hours of game footage and can recognize a vast array of patterns. They can discern the minute details of a player's movement, such as the angle of their stick or the speed of their dribble. This granular level of analysis was unimaginable a few years back but now is becoming a critical tool in a coach's arsenal.

Moreover, machine learning and deep learning technologies are invaluable for creating personalized training programs. Each player has unique strengths and areas for improvement. Traditional training programs often adopt a one-size-fits-all approach, but machine learning enables the customization of training modules based on individual performance data. Players can receive training that targets their specific weaknesses, whether that's defensive positioning or shooting accuracy. Personalized feedback accelerates skill development and helps each player unlock their full potential.

Beyond player improvement, these technologies revolutionize game strategies as well. Machine learning algorithms can analyze data from previous matches to predict future outcomes. They can consider various factors such as weather conditions, player fatigue, and even the psychological state of players to suggest the most effective strategies. Coaches no longer have to rely solely on their intuition or past experiences; they have data-driven insights that guide their decisions, increasing the chances of success on the field.

Performance analytics also undergo a transformation with machine learning and deep learning. These technologies can seamlessly integrate with wearable devices to monitor player metrics like heart rate, speed, and movement patterns in real-time. The result is a wealth of data that can be analyzed instantly to provide insights into player performance and health. Issues such as overtraining or the risk of injury can be identified early, allowing for timely interventions. This proactive approach helps maintain player health and ensures they are in optimal condition for matches.

Machine learning doesn't just stop at player improvement and game strategy; it also has applications in scouting and talent identification. Traditionally, scouts have had to rely on their keen eye for talent and experience, often traveling long distances to watch potential players in action. Machine learning can automate and enhance this process by analyzing game footage and performance data to identify promising players. Algorithms can evaluate not just physical skills but also game intelligence, decision-making abilities, and other intangibles that make for a successful field hockey player.

Deep learning, with its capability to process vast amounts of unstructured data, including videos and images, further refines the scouting process. Imagine a scout using an app that analyzes videos of young players from around the world, providing evaluations and even projecting future performance based on historical data. This global view of talent accelerates the identification and development of the next generation of field hockey stars.

While machine learning and deep learning bring an array of benefits, they also pose challenges. One such challenge is the need for high-quality data. The effectiveness of these technologies hinges on the availability of clean, accurate, and comprehensive datasets. Field hockey programs must invest in infrastructure to collect and manage this data effectively. Additionally, there's a learning curve for coaches and players to understand and trust these new tools. It requires a cultural shift within teams and organizations to embrace data-driven decision-making fully.

Nonetheless, the promise of machine learning and deep learning in field hockey is immense. They hold the potential to elevate the sport to new heights, driving innovations that create more dynamic, engaging, and intelligent gameplay. As these technologies continue to evolve, their integration into field hockey will become even more seamless and impactful, ushering in an era where data and intuition coexist to produce the best outcomes on and off the field.

In conclusion, machine learning and deep learning are more than just buzzwords; they are powerful tools reshaping the landscape of field hockey. From personalized training programs and enhanced game strategies to advanced scouting and real-time performance analytics, these technologies offer a comprehensive suite of solutions for everyone involved in the sport. Embracing these innovations will pave the way for a smarter, faster, and more efficient approach to playing and understanding field hockey.

AI in Other Sports

Artificial Intelligence (AI) isn't confined to a single sport; its applications span various athletic domains, each with unique demands and opportunities. While our primary focus revolves around field hockey, it's crucial to understand how AI integrates into other sports to fully appreciate its transformative potential. By examining AI's involvement in soccer, basketball, and even niche sports like archery, we glean insights that can subsequently be tailored to field hockey.

In soccer, AI has revolutionized several aspects of the game. Take, for example, sophisticated player tracking systems. These use machine learning algorithms to analyze player movements, offering insights that were previously unattainable. Coaches can review passing patterns, player stamina, and even tactical weaknesses. The data generated by AI can help create more efficient training programs, enhance tactical decisions, and ultimately improve team performance.

Think about basketball next. This sport has leveraged AI in fascinating ways, particularly through advanced video analytics. Automated cameras capture every move on the court, feeding data into AI systems that can break down each player's performance. By identifying subtle flaws in technique or positioning, coaches can provide highly specific feedback. This targeted approach makes practice sessions more effective and even reduces injury risks by spotting potential overuse vulnerabilities early.

Then there is American football, which has seen AI redefine scouting and game strategy. Teams use neural networks to evaluate thousands of potential plays, optimizing their strategies in real-time based on intricate data patterns. AI even helps in predicting opponent play-calling tendencies, thus providing a significant tactical edge. Football coaches and analysts rely on these AI systems to make informed, split-second decisions that can change the course of the game.

Don't overlook the more specialized sports like archery and golf. AI-based wearable devices have made significant strides in these sports, offering real-time feedback on an athlete's form and technique. For instance, in archery, sensors can detect minute deviations in an archer's stance or the bow's angle, providing immediate feedback to correct these errors. Similarly, in golf, AI-powered swing analyzers help players perfect their form by giving detailed insights into their swings, right down to the torque and angle of the club.

Baseball, with its exhaustive statistical background, naturally aligns with AI advancements. AI systems in baseball dive deep into metrics to offer predictive insights, from batting averages to pitch velocities. Performance metrics get dissected down to a granular level that would be impossible through manual analysis. These insights enable managers to make data-driven decisions about lineup configurations, pitching rotations, and even in-game strategies.

In the area of eSports, AI's role is just as transformative. The competitive gaming scene employs AI for multiple purposes, from game design to player training. AI can simulate thousands of game scenarios in a matter of seconds, allowing players to practice against computer-generated opponents with varying skill levels. This capability offers an invaluable edge in preparing for real-world competition. Moreover, AI analyses gameplay to identify patterns and strategies that could offer competitive advantages.

Moving towards motorsports, Formula 1 teams are utilizing AI to analyze colossal amounts of data generated during races. These include telemetry data, weather conditions, tire wear, and more. AI algorithms process this information in real-time, offering feedback to drivers and pit crews, who can make adjustments on the fly for optimal performance. The level of data analysis involved not only aids in strategy but also in the intricate designs of the cars themselves.

Ice hockey also benefits from AI implementations, particularly in goaltending and player tracking. High-speed cameras paired with machine learning algorithms can identify shot patterns, player speed, and even predict puck trajectories. This creates a more dynamic and informed approach to coaching and game strategy, as teams can adapt based on the real-time data provided.

On the tennis court, AI has stepped in to enhance both player performance and fan engagement. AI-based analytics tools study player strokes, movements, and match statistics to offer personalized training programs. These tools not only assist professional athletes but have trickled down to amateur levels, enabling broader application. Moreover, AI is used to generate player highlights in real-time during tournaments, thereby enriching the spectator experience.

In swimming, AI helps analyze the minutiae of strokes and turn techniques. Underwater cameras feed visual data into AI systems that can scrutinize every movement a swimmer makes. The detailed feedback helps swimmers refine their techniques to shave off crucial milliseconds from their performance times. Additionally, AI-based simulations can recreate different race conditions, allowing swimmers to train more effectively.

In track and field events, AI provides real-time analytics that are crucial for sprinters, hurdlers, and long-distance runners. High-speed cameras combined with AI can analyze the biomechanics of an athlete's run, offering insights that can improve efficiency and reduce the risk of injury. Coaches use this data to adjust training regimes, tailor individual workouts, and even fine-tune mental conditioning practices.

Another notable mention is cricket, a sport that combines elements of strategy, skill, and physical endurance. AI-driven systems analyze ball trajectories, player movements, and game strategies to offer comprehensive insights. These insights are crucial for coaches in making decisions on field placements, batting orders, and even in reviewing decisions made by umpires through technologies like Hawk-Eye.

In rugby, AI facilitates enhanced video analysis and player tracking, offering insights into tackle effectiveness, player fatigue, and even team synergy. These analytics aid coaches in devising better game plans and improving individual player performance. The real-time data from AI systems helps in making quick adjustments during the game, providing a clear advantage over teams that rely solely on traditional methods.

The landscape of sports is evolving rapidly, thanks to the integration of AI. While each sport presents unique challenges and opportunities, the underlying principles of AI application—data collection, predictive analysis, and real-time feedback—remain consistent. By examining AI's impact across these diverse sports, we can extract valuable lessons and strategies that can be employed to enhance the game of field hockey. This understanding not only equips players and coaches with advanced tools but also redefines the future possibilities in the world of sports.

Chapter 3:
AI in Field Hockey Training

Artificial intelligence is reshaping the landscape of field hockey training by offering a level of personalization and precision previously unimaginable. Personalized training programs, tailored to the unique strengths and weaknesses of each player, are now possible with AI-driven insights. This advanced technology evaluates a vast array of performance metrics to devise bespoke regimens that hone specific skills. Imagine a dynamic, AI-powered coaching assistant guiding players through drills that adapt in real-time to their progress, ensuring optimal development. Skill refinement tools further enhance players' techniques by providing instant feedback and suggesting minute adjustments. Whether it's perfecting a drag flick or improving defensive positioning, AI-driven coaching tools transform how athletes train, fostering an environment where continuous improvement becomes the norm. This leap forward doesn't just promise better individual performance; it builds smarter, more agile teams ready to compete at the highest levels.

personalized training programs

Incorporating AI into personalized training programs is transforming the realm of field hockey. By leveraging data-driven insights, AI can customize training regimens to cater to individual player's strengths and weaknesses. Gone are the days when a one-size-fits-all approach could be applied; today, the emphasis is on tailored training that optimizes player performance.

The beauty of AI lies in its ability to collect and analyze vast amounts of data quickly and efficiently. Every movement on the field, every pass and shot, and even players' physiological responses can be recorded and analyzed. From this analysis, AI can recommend specific drills and exercises that target areas needing improvement. For instance, a player struggling with penalty flicks can receive drills that focus on honing that particular skill with pinpoint precision, based on the AI's data analysis.

Moreover, wearable technology plays a crucial role in this personalized approach. Devices such as smartwatches, heart rate monitors, and GPS trackers feed real-time data into AI systems. This data ranges from biometric readings to movement patterns, providing a comprehensive look at each player's physical condition and performance metrics. As a result, coaches can construct more effective training schedules that improve fitness and skill while reducing the risk of injury. Individualized training plans can include specific fitness workouts, technical drills, and tactical exercises, all curated to meet the unique needs of each player.

This level of personalization extends beyond just drills and exercises. AI can also analyze psychological factors that impact performance. Mental toughness and game-day anxiety are critical components that can't be overlooked. By examining trends in performance under pressure, personalized mental conditioning programs can be designed. These plans might involve activities like mindfulness training, stress-relief exercises, and even cognitive-behavioral techniques aimed at boosting mental resilience.

Creating these personalized programs isn't a manual task. Advanced machine learning algorithms analyze data continuously, updating training plans as new data becomes available. This means that as a player's performance evolves, so does their training program. This dynamic adaptability ensures that the training remains relevant and effective, continuously pushing the player to their peak potential.

Another fascinating aspect of AI-driven personalized training is its ability to prevent overtraining and injuries. By monitoring players' physical loads, AI can predict when a player is at risk of injury due to fatigue or overexertion. This predictive capability allows coaches to adjust training intensity and volume, ensuring players maintain optimum performance levels without risking their health. This approach not only safeguards players' physical well-being but also enhances longevity in their careers.

Integrating AI into personalized training programs also fosters a more engaging and motivating training environment. Players receive instant feedback on their progress, highlighting areas of improvement and celebrating successes. This real-time feedback loop keeps them engaged and motivated to push their boundaries. The immediate recognition of improvements, no matter how small, can be incredibly empowering, encouraging players to persist in their efforts.

Team dynamics are another area where AI-designed personalized training shines. By understanding each player's unique profile, coaches can devise strategies that make the best use of individual strengths within the team's framework. AI can suggest the ideal roles and positions for each player based on their skills and performance metrics. This alignment maximizes both individual and team performance, ensuring that every player contributes effectively to the team's success.

Educational institutions and academies are also benefiting from these advancements. Young aspiring athletes can train more efficiently, developing their skills faster than previous generations. Tailored training programs, driven by AI, help these young talents to focus on the right areas from the start, creating a more efficient path to professional-level play. These institutions can also use AI to track and nurture the progress of their athletes over time, adjusting training programs as athletes grow and develop.

The integration of AI in personalized training programs is reshaping how players train, rest, recover, and perform. This transformation is evident at every level of the sport, from grassroots to professional leagues. As technology continues to evolve, the capabilities of these AI systems will only expand, offering even more precise and effective training solutions. The future of field hockey training is not just about hard work; it's about working smart, and AI is at the helm of this innovative shift.

While the impact of AI on personalized training is profound, it's essential to balance technology with human intuition. Coaches' experience and understanding of the game bring invaluable context that AI alone can't provide. The collaboration between AI-driven insights and human expertise creates a more holistic training experience, blending the best of both worlds. This synergy ensures that while data and algorithms guide the way, human touch remains ever-present, maintaining the soul of the sport.

Personalized training programs empowered by AI represent a significant leap forward in field hockey. By harnessing the power of data and advanced analytics, these programs offer a tailored, responsive, and highly effective approach to developing players. As AI continues to advance, its role in personalized training will only become more significant, promising an exciting future for players, coaches, and the sport itself.

Skill Development and Refinement

Artificial intelligence is revolutionizing skill development and refinement in field hockey. Gone are the days when training methods were solely based on a coach's intuition and player's self-assessment. AI brings precision, data, and science into the mix, resulting in significantly improved techniques and performance.

The first significant shift AI introduces in skill development is through data collection and analysis. Using advanced sensors and machine learning algorithms, AI systems can analyze a player's movements in real-time. These systems can track everything from the angle of a player's stick during a dribble to the force and trajectory of a shot. The data garnered is more precise than what the human eye can observe, allowing coaches to identify areas that need improvement with pinpoint accuracy.

Moreover, personalized feedback is a game-changer. Traditional coaching methods often apply a one-size-fits-all model, but AI can customize training programs to fit individual needs. For example, a player struggling with their drag flick technique can receive specific drills and exercises designed to target their weaknesses. Personalized training regimens not only make sessions more efficient but also enhance skill acquisition by addressing personal shortcomings directly.

Furthermore, AI-driven tools can simulate game scenarios, offering players a chance to refine their skills in a variety of contexts. These simulations can replicate high-pressure situations, like taking a penalty corner in the last minute of a tied game, enhancing a player's ability to perform under stress. For instance, virtual reality systems can immerse players in lifelike match environments, enabling them to practice and refine their skills with a level of engagement previously unattainable.

Another critical aspect of AI in skill refinement is its ability to track progression over time. Regular assessments using AI-tools provide a comprehensive view of a player's development. Visualizing this data helps both players and coaches understand what's working and what's not, making continual adjustments to the training regimen. This dynamic feedback loop accelerates learning and ensures that players are always working toward meaningful improvements.

In addition to physical skills, AI also aids in developing cognitive skills. Decision-making, spatial awareness, and game intelligence are crucial aspects of field hockey. AI tools can analyze game footage and highlight moments where a player made a suboptimal decision. By reviewing these moments, players can better understand their thought processes and work on making quicker, more effective decisions during matches.

Team play is another area where AI makes significant contributions. AI can analyze how well players coordinate with each other, identifying patterns and suggesting improvements. These insights help teams enhance their on-field chemistry, leading to smoother, more coherent play. Coaches can also use these analyses to develop tailored drills that improve teamwork and strategic execution.

Skill development isn't confined to young players or amateurs. Even seasoned professionals can benefit from AI's insights. As players age, their physical abilities may decline, but with AI, they can adapt their play style to maintain a high level of performance. For example, a veteran player can shift focus from speed to positioning, using AI to fine-tune their movements and maintain their competitive edge.

Lastly, the power of AI isn't limited to individual training sessions. During team practices, AI can provide real-time feedback, allowing for immediate corrections. Imagine a scenario where during a scrimmage, an AI system alerts a defender about their positioning or advises a midfielder on optimal passing options. This immediate feedback helps players make adjustments on the fly, ingraining better habits faster.

Additionally, the refinement process isn't only about correcting errors but also about reinforcing strengths. AI can highlight what a player does well, ensuring these strengths become even more robust. Building confidence through positive reinforcement can be just as critical as fixing flaws, making players more well-rounded and resilient.

In the scope of skill development, the intersection of AI and field hockey is just the beginning. As technology evolves, the tools and methods for refining skills will only become more advanced, pushing the boundaries of what players can achieve. The relentless pursuit of excellence, aided by state-of-the-art AI, promises an exciting future for field hockey enthusiasts everywhere.

Fundamentally, the integration of AI in skill development and refinement offers a comprehensive and multi-faceted approach to training. By leveraging data, personalized feedback, cognitive and physical simulations, and real-time analytics, this technology cultivates a new breed of athletes who are more adaptable, skilled, and mentally prepared than ever before. As we stand on the precipice of this new era in sports training, the potential for growth and excellence seems virtually limitless.

To encapsulate, AI's contribution to skill development in field hockey exemplifies a perfect blend of high-tech innovation and athletic mastery. The continually evolving tools not only sharpen individual skills but foster better teamwork and smarter gameplay, driving the sport forward into a future of unprecedented performance levels and strategic sophistication. The marriage of AI and field hockey is more than just a technological advancement; it's a revolution in how the game is played, trained, and understood.

AI-Driven Coaching Tools

Artificial intelligence is not merely a technological buzzword; it's become a game-changer in the athletic world, including field hockey. AI-driven coaching tools are revolutionizing how coaches approach training, game strategy, and player development. These tools offer a treasure trove of actionable insights, automated analyses, and personalized feedback that were unimaginable just a decade ago. They are reshaping coaching methodologies, providing a competitive edge that was once the realm of intuition and experience. In this section, we'll explore how AI-driven coaching tools are making waves in the field hockey training ecosystem.

In traditional coaching, the emphasis has often been on the coach's ability to read the game and understand the players' strengths and weaknesses. It relies heavily on observations, experience, and gut feeling. While these elements are indispensable, AI-driven coaching tools introduce a new dimension by adding precision and data-driven insights to the coach's arsenal. These tools leverage machine learning algorithms to analyze myriad aspects of the game, including player performance, tactics, and even psychological factors.

One of the most impactful areas where AI-driven tools shine is in personalized training programs. By collecting and analyzing data from various sources—such as wearables, video footage, and performance metrics—AI can recommend customized training regimens tailored to each player's unique needs. For instance, if a player's sprint speed is lagging, the AI tool can suggest specific drills to enhance this skill. This approach ensures that every minute of training is spent effectively, targeting areas that need improvement while maintaining strengths.

Beyond individual development, these tools also enable coaches to refine team dynamics. AI can study player interactions, positioning, and movement patterns in real-time, offering insights that might be missed through human observation alone. For instance, AI can highlight subtle inefficiencies in team formations or coordination, allowing the coach to adjust strategies on the fly. This level of analysis helps to create a more cohesive unit, ultimately leading to improved performance on the field.

AI-driven tools also focus heavily on skill refinement. Advanced video analysis powered by computer vision technology allows coaches to break down complex movements frame by frame. This granular level of detail enables a more thorough assessment of techniques like dribbling, passing, and shooting. Players can then receive instant feedback, complete with visual aids and corrective measures. Such precise feedback loops accelerate learning and mastery, giving athletes a significant advantage.

One intriguing application of AI in coaching is the use of predictive analytics to anticipate future performance and game outcomes. By analyzing historical performance data alongside current metrics, AI-driven tools can forecast trends and potential issues before they become problematic. This predictive capacity allows coaches to make more informed decisions, such as when to rest a player or implement new tactics. It's akin to having a crystal ball, but backed by rigorous data and analysis.

The interaction between AI and human coaching is not merely additive; it's symbiotic. While AI can process vast amounts of data at lightning speed, the coach provides the contextual understanding and emotional intelligence necessary to interpret these insights meaningfully. This partnership helps to create a holistic coaching approach that leverages the best of both worlds. Coaches can focus more on strategic and psychological aspects, knowing that the AI is handling the heavy lifting of data analysis.

Moreover, AI-driven tools are democratizing access to elite-level coaching insights. Historically, such detailed analytics were the purview of well-funded professional teams. However, advances in technology are making these tools more accessible to amateur and youth programs, leveling the playing field. This democratization empowers coaches at all levels to make data-driven decisions, fostering a more inclusive and competitive environment for talent development.

However, the adoption of AI-driven coaching tools is not without its challenges. Coaches need adequate training to effectively interpret and use the data generated by these tools. Education programs focusing on AI literacy in sports are essential to bridge this gap.

Additionally, the cost and complexity of implementing these systems can be daunting. Teams must weigh the investment against the potential benefits and devise strategies to integrate these tools seamlessly into their existing workflows.

As the technology continues to evolve, the potential applications of AI-driven coaching tools will only expand. Future innovations could include real-time decision support systems, virtual reality (VR) training environments, and even AI-generated scouting reports. These advancements promise to push the boundaries of what's possible in field hockey coaching, offering unprecedented levels of insight and efficiency.

In conclusion, AI-driven coaching tools are not just enhancing how training is conducted; they are transforming the very philosophy of coaching in field hockey. By combining the analytical power of AI with the nuanced expertise of human coaches, teams can achieve higher performance levels and sustained success. As AI continues to advance, its role in field hockey training will only grow, making it an indispensable tool in the coach's toolkit. This symbiosis between technology and human intuition heralds a new era in sports coaching, promising to redefine the game for future generations.

Chapter 4:
Enhancing Game Strategy with AI

As we delve into enhancing game strategy with AI, it's clear that artificial intelligence is revolutionizing how teams and coaches approach tactical decisions in field hockey. The integration of real-time data analytics, combined with predictive modeling, empowers teams to make split-second decisions that can alter the course of a match. By analyzing vast amounts of data, AI identifies patterns and trends that are often invisible to the human eye, offering a strategic edge that was previously unimaginable. AI-assisted decision-making tools can simulate numerous game scenarios, providing coaches with actionable insights to refine their game plans and tactical maneuvers. This blend of technology and athletic prowess not only diversifies the strategic options available but also elevates the game to a new level of sophistication and competitiveness, creating a dynamic and predictive environment where every move is calculated for maximum impact.

Real-Time Data Analytics

As field hockey transitions into the realm of precision and tactical finesse, the role of real-time data analytics becomes indispensable. Imagine a game where coaches and players have access to granular insights in the blink of an eye, making it possible to alter strategies on the fly. This is not a distant fantasy; it's a present-day reality, thanks to advancements in AI and real-time data analytics.

Initially, the concept of utilizing real-time data during matches can seem overwhelming. However, breaking it down reveals numerous tangible benefits. With the deployment of sensors and advanced tracking systems, data can be collected instantaneously and relayed to coaching staff, players on the pitch, or even analysts seated miles away. This data can encompass everything from player positioning and ball trajectory to player fatigue levels and likelihood of injury. Such a robust system transforms field hockey into a game of not just skill but also strategic depth informed by rich data streams.

At the core of real-time data analytics is the ability to collect data accurately and quickly. This involves the integration of technologies such as GPS trackers, inertial measurement units (IMUs), and advanced cameras. These tools work in tandem to capture real-time metrics like speed, distance, acceleration, and heart rate, providing a comprehensive view of a player's physical and tactical performance at any given moment. These metrics form the basis for nuanced analysis that can drive immediate in-game adjustments.

One of the essential benefits of real-time data analytics is the ability to map out player movements and patterns in real-time. By visualizing data, coaches can observe how well team formations are holding up, identify gaps in defense, or exploit weaknesses in the opposing team. For instance, if the data indicates a particular player is consistently out of position, the coach can address this immediately, rather than waiting for a post-game analysis. This kind of dynamic intervention can be the difference between winning and losing.

Real-time data isn't just for the benefit of the coaching staff. Players themselves can gain insights on the go. Wearable devices can display vital information directly to players, helping them manage their exertion levels and make more informed decisions. Imagine a scenario where a forward player is informed through his wearable tech that he's approaching optimal fatigue levels. This immediate feedback enables him to adjust his gameplay, conserving energy for critical moments when a burst of speed might be necessary.

Consider the tactical advantage when analyzing your opponent in real-time. If data reveals that the opposing team's midfielder is their pivot player, the strategy can be promptly adjusted to neutralize that threat. Alternatively, detecting an opponent's declining performance metrics offers an opportunity to exploit that weakness. Swift reconfiguration of player roles or formations based on live data can create unexpected advantages, putting your team in the driver's seat.

There's also a psychological component to the use of real-time data. Athletes often describe the experience of "being in the zone," a state where they perform at their peak without conscious effort. Real-time data analytics can cultivate this state by reducing uncertainty and allowing players to focus on the game, rather than being bogged down by decisions. When uncertainty is minimized, athletes are more likely to experience flow, enhancing overall team performance.

The utility of real-time data extends beyond immediate game situations. In training sessions, the live feedback can simulate game conditions, making practice more effective. Players can test different strategies and immediately see what works and what doesn't, fostering a learning environment driven by data-informed insights. This continuous feedback loop enhances skill development and strategic acumen.

From a tactical perspective, the immediate feedback transforms traditional halftime adjustments into a continuous process. Coaches no longer have to wait until halftime to provide input; they can make these adjustments during the game. These micro-adjustments keep the team aligned and responsive to ever-changing game scenarios.

Real-time data analytics also offers a comprehensive way to measure the effectiveness of these in-game strategies. Each change made based on real-time data presents an opportunity to track its impact, giving coaches and analysts the ability to fine-tune their approach. Over time, this results in a well-honed strategy, backed by empirical evidence rather than intuition alone.

However, integrating real-time data analytics into field hockey isn't without challenges. The volume of data generated can be immense, requiring sophisticated algorithms to process and interpret it in meaningful ways. The efficacy of real-time data analytics hinges on the quality of the AI models employed. These models must be trained on vast datasets to ensure they can accurately predict and identify critical gameplay elements.

Moreover, with this wealth of data, safeguarding sensitive information becomes paramount. Field hockey organizations must prioritize data security to protect both competitive secrets and athlete privacy. Ethical considerations must guide how this data is collected, stored, and used, ensuring that it serves to enhance the sport without compromising the integrity or well-being of the players.

Finally, it is crucial to understand that while real-time data analytics provides a powerful tool, the human element remains irreplaceable. Coaches' experience and intuition, combined with robust data, create a synergy that can elevate team performance to new heights. Players' instincts and creativity on the pitch are equally indispensable in leveraging data effectively.

In conclusion, real-time data analytics represents a transformative force in the landscape of field hockey. By providing actionable insights at breakneck speed, it allows teams to adapt and thrive in the heat of the moment. This marriage of technology and human ingenuity is not just the future of field hockey; it's the bedrock of today's most successful teams. As AI and data analytics continue to evolve, the potential for deeper and more sophisticated integration will only grow, promising to push the boundaries of what's possible on the field.

Predictive Modeling for Game Strategy

As we delve deeper into the realm of predictive modeling for game strategy, it's vital to recognize that field hockey, like many other sports, thrives on the ability to anticipate and adapt. In a game where split-second decisions can determine the outcome, the capability to forecast events and react proactively offers a competitive edge that cannot be understated.

Predictive modeling employs a combination of historical data, machine learning algorithms, and statistical techniques to project future occurrences based on past patterns. For field hockey, this means identifying trends in player movements, understanding opponent strategies, and even predicting potential injuries. What makes this remarkable is the shift from mere reactive strategies to anticipatory, proactive maneuvers.

One of the fundamental aspects of predictive modeling in field hockey revolves around data acquisition. Modern field hockey is inundated with data points—from player positioning to ball possession metrics. The use of advanced sensors, video analysis, and GPS technology has made it possible to gather and analyze vast amounts of data in real-time. These data points serve as the raw material for predictive models, enabling a detailed understanding of game dynamics and individual player performance.

At the heart of predictive modeling lies the power to simulate various game scenarios. By feeding the model with different sets of input parameters, coaches and analysts can forecast outcomes based on different strategies. For instance, how might a team's defensive structure hold up against a particular opponent's attacking formation? By running these simulations, teams can identify optimal strategies and prepare contingency plans.

Imagine a scenario where a team is preparing for a crucial match against a league leader. Instead of relying on gut instincts and historical matchups alone, the coaching staff can leverage predictive models to analyze the opponent's weaknesses. These models can highlight patterns in the opposing team's play—such as a predictability in their attack during the final quarter or susceptibility to counterattacks when pressing high. Armed with this knowledge, the coaching staff can devise a finely-tuned strategy to exploit these weaknesses and potentially turn the tide in their favor.

While predictive modeling offers incredible advantages, it doesn't come without challenges. One of the significant hurdles is the quality and granularity of data. For the models to be accurate, they require high-quality, detailed data, which can sometimes be difficult to obtain consistently. Moreover, the models must be continuously updated with fresh data to remain relevant and accurate. The dynamic nature of sports means that past data can quickly become outdated if not regularly refreshed.

Another challenge lies in the complexity of human behavior. Players aren't robots; their decisions can be influenced by numerous factors including mental state, fatigue, and team dynamics. Predictive models must incorporate these variables to provide a holistic and accurate prediction. This often requires advanced machine learning algorithms and a multidimensional approach to data analysis.

Despite these challenges, the benefits of predictive modeling in field hockey are manifold. Teams can manage their player rotations more effectively, ensuring that key players are rested and in peak condition for crucial moments. Injury prevention can be enhanced by predicting which players are at higher risk based on their workload, playing style, and historical injury data. By anticipating these risks, teams can take preemptive measures to reduce injuries and keep their squad healthy throughout the season.

Moreover, predictive modeling extends beyond just on-field strategies. It can be instrumental in talent scouting and development. By analyzing a young player's performance data, teams can predict their potential career trajectory and development needs. This helps in making informed decisions during talent acquisition and ensures that resources are allocated efficiently to nurture future stars.

Ultimately, predictive modeling is reshaping the landscape of field hockey by embedding a scientific and data-driven approach to strategy. The ability to foresee events and adapt accordingly offers a strategic advantage that is becoming increasingly crucial in the modern game. As teams continue to leverage these advanced technologies, we can expect to see a more intricate and competitive field hockey environment, where the margins for error are slimmer and the stakes are higher.

This transformation is a testament to the ongoing evolution of the sport, driven by the relentless pursuit of excellence and innovation. By integrating predictive modeling into their strategic arsenal, field hockey teams are not just reacting to the game; they are actively shaping it, turning the unpredictable nature of sport into a calculated and tactical endeavor.

In the end, the true power of predictive modeling lies in its ability to enhance human decision-making. It doesn't replace the instinct and experience of seasoned coaches and players, but rather, augments it with precise and actionable insights. This synergy between human intuition and artificial intelligence is where the future of field hockey strategy lies, promising a new era of smart, efficient, and exhilarating gameplay.

AI-Assisted Decision Making

Field hockey, like many other sports, is rapidly evolving with the advent of artificial intelligence (AI). Among the myriad ways AI is being integrated into the sport, AI-assisted decision making stands out as transformative, particularly in refining strategic approaches and in-game tactics. Coaches and analysts are finding themselves equipped with tools that can dissect the game at a granular level, offering insights that were previously unattainable.

At the heart of AI-assisted decision making lies real-time data analysis. During a match, countless data points are generated—player positioning, ball movement, passing patterns, and even individual player fatigue levels. AI algorithms process this data swiftly, identifying patterns and anomalies that may influence the game's outcome. This allows coaches to make informed decisions on the fly, such as adjusting formations or substituting players at crucial moments.

One significant advantage of AI is its predictive capabilities. Utilizing machine learning models, AI can analyze historical data to predict future events with remarkable accuracy. For instance, by studying an opponent's past games, AI can forecast their likely strategies and suggest countermeasures. This kind of foresight gives teams a competitive edge, allowing them to prepare for various scenarios before they unfold on the field.

AI's role is not limited to just the matches. It also extends to training sessions where it provides prescriptive analyses. During practice, AI can suggest which drills players need based on their performance metrics, ensuring that each session is tailored to address specific weaknesses and enhance strengths. Over time, this personalized approach leads to a more cohesive and well-prepared team.

If we dive deeper, AI can aid in refining individual player performance. By analyzing video footage and tracking detailed player movements, AI can provide insights into aspects like stick handling, shot accuracy, and defensive positioning. These insights are not merely observational; they come with actionable recommendations on how to improve. For example, slight tweaks in a player's stance or movement might be suggested to enhance efficiency and effectiveness on the field.

The tactical elements of field hockey can be highly complex, and AI-assisted decision making simplifies this complexity. Take, for instance, the task of optimizing team formations. Traditionally, coaches would rely on intuition and experience to choose formations, but AI can process vast amounts of data to determine the most effective setups against specific opponents. This means formations can be dynamically adjusted to exploit an opponent's weaknesses or to strengthen defensive capabilities.

Another area where AI shines is in managing player workload to prevent burnout and injuries. Continuous monitoring of physical metrics during training and matches allows AI to identify signs of fatigue and recommend rest or adjusted training loads. This proactive approach helps maintain player health and ensures peak performance during critical periods.

It is crucial to recognize that AI's role in decision making is not to replace human intuition and experience but to complement them. AI provides a data-driven foundation upon which coaches can build their strategies. The blend of AI analysis and human expertise creates a powerful synergy, leading to smarter and more strategic decisions that can alter the course of a game.

A real-world example of AI-assisted decision making can be found in penalty corner situations, one of the most critical and pressure-filled moments in field hockey. AI can analyze endless combinations of offensive and defensive setups to determine the most effective strategies based on the current game's context and the capabilities of the players on the field. This level of detailed preparation helps teams execute their plans with precision, often making the difference between victory and defeat.

Moreover, AI can democratize high-level strategic insights. Smaller teams, which may not have access to top-notch coaching staff or vast analytical resources, can leverage AI tools to gain a strategic edge. This levels the playing field, making the sport more competitive and exciting.

The integration of AI in decision making also promotes continuous learning and adaptation. Post-game analyses generated by AI offer insights that help teams understand what worked, what didn't, and why. This feedback loop enables constant strategy refinement, ensuring that both coaches and players are continuously evolving and improving.

The exciting prospect of AI-assisted decision making in field hockey is its potential to foster innovation. As AI continues to evolve, it will undoubtedly bring new ways to interpret data and generate insights. The key for teams will be to stay agile and embrace these innovations to stay ahead in the game.

While we've discussed many benefits, it's essential to address some challenges, too. Data quality and accuracy are paramount; if the data fed into AI systems is flawed or incomplete, the recommendations and decisions derived from that data may not be reliable. Hence, maintaining data integrity through rigorous validation and cleaning processes is a critical aspect of utilizing AI effectively.

Furthermore, the ethical implications of AI-assisted decision making cannot be ignored. It's vital to ensure transparency in how AI models make decisions and to guard against any biases that might be encoded in the algorithms. Stakeholders within field hockey must work collaboratively to establish norms and guidelines that ensure the responsible use of AI.

Despite these challenges, the trajectory of AI in enhancing game strategy is overwhelmingly positive. The combination of instant, data-backed insights and human ingenuity is transforming how decisions are made in field hockey. As AI technology continues to advance, it is poised to unlock even greater strategic capabilities, making the sport more dynamic and strategically complex.

In conclusion, AI-assisted decision making is reshaping the landscape of field hockey strategy. With real-time analytics, predictive modeling, personalized training, and enhanced player management, AI offers a suite of tools that empower teams to make smarter, more informed decisions. The integration of AI in this domain is not just a technological upgrade; it's a strategic revolution that benefits coaches, players, and analysts alike. This, indeed, is a thrilling era for field hockey, marked by the fusion of tradition and cutting-edge technology.

Chapter 5:
Performance Analysis and Metrics

Performance analysis and metrics in field hockey have taken a quantum leap with the advent of AI, fundamentally transforming how players, coaches, and analysts approach the game. Gone are the days of relying solely on human intuition and rudimentary stats; AI-driven tools now dissect every nuance of a player's performance. Key performance indicators (KPIs) offer an in-depth look at a variety of critical metrics, helping to identify strengths and areas for improvement. Personalized feedback, derived from sophisticated algorithms, helps tailor training regimes to individual athlete's needs. Moreover, video analysis supercharged by AI can break down play sequences frame by frame, offering actionable insights almost in real time. These advancements make it clear: understanding and leveraging performance analysis and metrics through AI is no longer just a supplemental advantage but an essential aspect of modern field hockey.

Key Performance Indicators in Field Hockey

Field hockey, like any competitive sport, thrives on measurable outcomes. These outcomes, often referred to as Key Performance Indicators (KPIs), are vital for understanding a player's or team's performance. KPIs provide a quantifiable measure of success and development, serving as the foundation for performance analysis in field hockey.

The realm of KPIs in field hockey encompasses a wide range of metrics. Some are directly tied to the game's fundamental aspects, such as goals scored, assists, and saves. Others delve deeper, analyzing the minutiae of actions on the field. Each KPI offers unique insights, helping coaches and players focus on specific areas that need improvement or celebrate areas of strength.

Goals, assists, and saves are perhaps the most straightforward KPIs. These provide a clear snapshot of offensive and defensive prowess. However, field hockey is much more nuanced, and relying solely on these metrics could be misleading. Hence, it becomes essential to look at advanced metrics that paint a more comprehensive picture.

Pass completion rate is one such advanced metric. This KPI measures the accuracy of a player's passes, shedding light on their ability to maintain possession and contribute to the team's overall fluidity. A high pass completion rate often correlates with effective ball control and strategic decision-making on the field. Conversely, a low rate may indicate rushed decisions or a need for skill refinement.

In addition to pass completion, distance covered during a game is a crucial KPI. This metric helps in understanding a player's work rate and endurance. Field hockey demands extensive movement, and covering vast distances efficiently can indicate a player's physical fitness and their understanding of tactical positioning.

Take, for example, a midfielder who consistently covers more ground than their teammates. This player may not always be directly involved in critical moments but plays a pivotal role in maintaining team shape and transitioning between defense and attack. Monitoring distance covered helps in planning personalized training programs aimed at enhancing a player's stamina and tactical awareness.

A related KPI is the number of sprints executed during a match. Field hockey is a game of bursts, requiring frequent sprinting interspersed with periods of lighter activity. Tracking sprint frequency and intensity offers insights into a player's explosive power and their ability to respond quickly to the changing dynamics of the game.

Defensively, successful tackles and interceptions are key metrics. These KPIs show a player's ability to disrupt the opposition's plays and regain possession. A defender who consistently notches high numbers in these areas is invaluable, contributing to a team's overall defensive solidity.

Analyzing these defensive metrics goes beyond just numbers. It involves examining the context of each tackle or interception to understand the player's decision-making process. This analysis can reveal patterns, such as a tendency to dive into tackles recklessly or an exceptional ability to read the game and anticipate opponent movements.

Shot accuracy and penalty corner conversion rates are critical for evaluating offensive efficiency. Shot accuracy measures the percentage of shots on target, providing a clear picture of a player's precision under pressure. Similarly, penalty corner conversion assesses a team's proficiency in exploiting set-piece opportunities.

Field hockey teams often invest considerable time in perfecting penalty corners, given their potential impact on the game's outcome. Tracking conversion rates helps in identifying the effectiveness of various corner strategies and the execution skills of both attackers and drag-flickers.

Possession metrics are another essential aspect of performance analysis. Maintaining possession not only reduces the opponent's opportunities but also allows a team to dictate the game's tempo. Possession percentage and the number of turnovers provide a snapshot of a team's control over the game. High possession percentages typically signal dominance, while frequent turnovers indicate areas needing improvement in ball handling or decision-making under pressure.

A well-rounded analysis would be incomplete without considering fouls and cards. These negative KPIs inform about a team's discipline and adherence to the rules. Frequent infractions can disrupt team dynamics, lead to numerical disadvantages, and signify underlying issues with player behavior or defensive techniques.

By codifying these KPIs and making them accessible through advanced AI tools, coaches, and analysts can gain deeper insights into performance trends. This systematic approach allows for accurate assessments, paving the way for targeted improvements. Furthermore, by combining these metrics, AI can offer predictive analytics, helping teams anticipate future performance based on past data.

Incorporating AI in performance analysis transforms these KPIs from static numbers to dynamic insights. Machine learning algorithms can process vast amounts of data to detect patterns and correlations that might escape the human eye. For instance, AI can identify which combinations of players perform best together, or how specific plays impact overall performance. These insights can inform game strategy and training focus areas, leading to a more holistic approach to performance enhancement.

However, the real magic lies in contextual interpretation. Numbers alone don't tell the whole story; understanding the context behind each KPI is crucial. Whether it's the timing of an assist or the situational pressure during a crucial save, these nuances can only be captured through comprehensive analysis. AI can play a significant role in this contextual interpretation by integrating multiple data sources and providing a layered understanding.

Ultimately, the goal of tracking KPIs in field hockey is not just to highlight areas of weakness but to celebrate areas of strength. These indicators provide a roadmap for continuous improvement, guiding players and teams on their journey to excellence. With AI's help, this journey becomes more data-driven, allowing for precise interventions and fostering a culture of constant evolution.

The marriage of technological sophistication and human intuition is key to unlocking field hockey's full potential. As we delve deeper into AI's role in this sport, it becomes evident that KPIs are more than mere numbers. They are the building blocks of a strategy, the milestones of progress, and the harbingers of future victories. The future of field hockey is bright, with KPIs serving as guiding stars, illuminating the path to unparalleled performance.

AI Tools for Performance Tracking

Field hockey, like many sports, is increasingly turning to advanced technologies to improve performance and refine strategies. AI tools for performance tracking stand at the forefront of this tech revolution, offering unprecedented insights into player metrics and game dynamics. By harnessing the power of AI, coaches and analysts can dive deep into data, which was once only available through extensive manual labor, often hampered by subjective biases. Now, real-time analytics can provide factual, timely, and actionable insights.

One of the primary advantages of AI tools is their ability to process vast amounts of data quickly and accurately. For example, wearable devices equipped with AI can monitor players' physiological metrics such as heart rate, speed, and distance covered during training and matches. These wearables provide raw data that AI algorithms analyze to generate detailed reports on a player's fitness levels, performance, and areas that need improvement. These insights enable coaches to customize training regimens tailored to each player's specific needs, optimizing their performance on the field.

Performance tracking tools aren't just confined to wearable technology. AI-driven software solutions can analyze video footage to provide comprehensive break-downs of matches. Video analysis powered by AI can automatically identify patterns in gameplay, player movements, and team formations. For instance, computer vision algorithms can track player positioning, ball movements, and goal opportunities with remarkable precision. These tools not only save time but also eliminate the errors associated with manual tracking.

Beyond the immediate benefits to training and gameplay, these AI systems contribute to long-term performance tracking as well. By compiling historical data, they help in creating performance baselines for both individual players and teams. Trends can be identified, allowing for the anticipation of performance dips or the early identification of rising stars. This longitudinal data becomes vital for career development, injury prevention, and even psychological preparedness.

The integration of AI tools for performance tracking also extends to tactical analyses. AI can dissect an opponent's strategies and uncover weaknesses and patterns. Teams can thus enter matches better prepared, having tailored their strategies to exploit identified gaps in the opposing team's defenses. This kind of intelligence was previously the purview of intuitive, experienced coaches; AI democratizes access to such high-level insights.

Moreover, AI tools relate seamlessly with other performance metrics that have long been essential in field hockey, offering multi-dimensional analyses. Metrics like pass accuracy, interception rates, shooting efficiency, and defensive coverage are all dissected with an analytical rigor that enhances traditional methods. Players receive real-time feedback, coaches refine strategies on the fly, and analysts provide deeper insights during post-game reviews.

While these technologies are undeniably transformative, their application must be sensibly managed. The massive influx of data can be overwhelming, and there is always a risk of over-reliance on AI outputs. Human intuition, experience, and judgment remain indispensable parts of the equation. AI tools should serve as enhancements, not replacements, of the insights that skilled coaches and players bring to the game.

Additionally, the ethical considerations linked to AI tools for performance tracking shouldn't be overlooked. Issues concerning data privacy and the potential misuse of sensitive information must be rigorously addressed. Players should have clarity on how their data is collected, stored, and utilized. Ensuring transparency and building trust among all stakeholders is paramount for the successful integration of AI in performance tracking.

Ultimately, AI tools for performance tracking offer a fusion of precision, efficiency, and insight that was previously unattainable. Their capability to deliver granular, real-time data, comprehensive video analyses, and actionable insights gives coaches and players an edge in a highly competitive environment. Realizing the full potential of these technologies requires a balanced approach, informed by both technical prowess and ethical responsibility. Yet, there is no denying that AI is transforming the landscape of field hockey, pushing the boundaries of what can be achieved on the field.

Video Analysis and AI

Video analysis and artificial intelligence are revolutionizing performance analysis and metrics in field hockey, providing an unprecedented level of detail and insight into every aspect of the game. By leveraging AI, coaches and analysts can now dissect game footage with an accuracy and depth that was previously unimaginable. This fusion of technology empowers teams to uncover strengths, identify weaknesses, and fine-tune strategies in a way that is both efficient and highly effective.

Central to this transformation is the use of video analysis software, which employs computer vision and machine learning algorithms to identify and track players, movements, and even the ball. These systems can automatically generate detailed reports on player positioning, passing accuracy, and defensive formations, among other metrics. For instance, advanced optical tracking solutions can capture minute details such as player velocity, acceleration, and collision impacts, providing a comprehensive view of on-field dynamics.

The integration of AI in video analysis isn't just about crunching numbers—it's about generating actionable insights. For example, video feeds can be analyzed in real-time during games, providing coaches with immediate feedback. This allows for in-game adjustments that can shift the momentum in favor of the analyzing team. Think about a coach recognizing a gap in the opponent's defense through AI-driven video analysis and swiftly adapting the team's offensive strategy to exploit it.

AI's role in video analysis extends beyond live games to practice sessions as well. By analyzing training footage, AI can evaluate individual player performance, track improvement over time, and suggest personalized drills. For example, if a player's dribbling accuracy needs work, the system can identify specific moments where errors are most frequently occurring and recommend targeted exercises. This personalized feedback loop helps in honing skills more effectively compared to traditional methods.

Another exciting development is the use of predictive analytics in video analysis. By examining historical footage along with current game data, AI models can forecast game outcomes and player performance trends. This predictive power can be invaluable for strategic planning. Teams can prepare for matches with a data-driven understanding of both their own performance and that of their opponents, adjusting strategies based on predicted scenarios. For example, a model might highlight that an opposing team tends to falter in the second half when facing high-pressure tactics, allowing a coach to devise a strategy that capitalizes on this insight.

Video analysis enhanced by AI also democratizes access to elite-level coaching tools. In the past, such in-depth analysis was available only to top-tier teams with extensive resources. Now, AI-powered video analysis tools can be utilized by any team with a standard or mobile camera setup. This accessibility enables youth teams, amateur clubs, and schools to benefit from the same cutting-edge technologies that professional teams use, leveling the playing field and advancing the sport as a whole.

Moreover, AI video analysis is redefining the concept of performance metrics in field hockey. Traditional metrics like goals scored and assists recorded still hold significance, but AI introduces advanced metrics such as expected goals (xG), passes leading to shot attempts (key passes), and possession value. These next-generation metrics provide a more nuanced understanding of a player's contribution and tactical impact, allowing coaches to make more informed decisions regarding lineup selections, in-game tactics, and player development.

Another significant advantage of AI in video analysis is its ability to minimize human error and bias. Objective algorithms consistently analyze plays without preconceived notions, ensuring unbiased assessments of player performance. This objectivity is particularly useful in contentious situations where human judgment may be swayed by emotion or preconceived biases. With AI, every decision and analysis is based on data and empirical evidence, promoting fairness and transparency in performance evaluations.

The application of AI in video analysis is continuously evolving, leveraging advancements in technology and data science. Deep learning, for example, has made it possible for machines to learn from vast amounts of data and improve their analytical capabilities over time. With each analyzed game or training session, AI models become more adept at recognizing patterns, detecting anomalies, and providing actionable insights, resulting in a dynamic and constantly improving analytical tool.

It's not just the coaches and analysts who benefit from these advancements. Players, too, are gaining a deeper understanding of their performance through AI-driven video analysis. By reviewing personalized video clips and detailed performance reports, players can visualize their strengths and weaknesses, leading to more targeted and effective training. For example, a defender might learn about their positioning errors during counterattacks, while a forward might identify patterns in their movement that lead to successful goal-scoring opportunities.

Importantly, the insights gained from AI video analysis have far-reaching implications for injury prevention and player longevity. By monitoring player workload, movement patterns, and biomechanical data, AI can identify potential injury risks long before they materialize into actual injuries. This proactive approach allows for timely interventions, such as adjusting training loads, incorporating specific rehabilitation exercises, or even altering playing styles to mitigate injury risks, ultimately enhancing player safety and career longevity.

The potential of video analysis and AI in field hockey is vast and continues to grow as technology advances. The combination of high-definition video footage, sophisticated AI algorithms, and vast amounts of data creates a powerful tool for optimizing performance, enhancing strategic planning, and fostering player development. As these technologies continue to evolve, their impact on the sport will only grow, shaping the future of field hockey in profound ways.

Adopting AI-based video analysis requires a cultural shift within teams and organizations, embracing data-driven decision-making and continuous learning. Coaches, players, and analysts must be willing to integrate these technologies into their daily routines and trust the insights generated by AI. To facilitate this transition, ongoing education and training are essential, ensuring that everyone involved understands the capabilities and limitations of AI-driven video analysis and can effectively leverage its potential.

In conclusion, AI's integration into video analysis is reshaping the landscape of performance analysis and metrics in field hockey. By providing detailed, objective, and actionable insights, AI-powered video analysis enables teams to refine strategies, enhance player performance, and ultimately achieve greater success on the field. As the technology continues to advance, embracing AI-driven video analysis will be crucial for staying competitive in the fast-evolving world of field hockey, unlocking new levels of performance and innovation in the sport.

Chapter 6:
Injury Prevention and Management

The seamless integration of artificial intelligence into injury prevention and management is transforming how field hockey players maintain peak performance while reducing downtime. Wearable technology, such as smart clothing and advanced sensors, provides real-time data on players' biomechanics and physiological states, enabling early detection of potential injuries. Predictive injury models analyze this data to forecast injury risks, allowing for proactive rehabilitation and personalized training adjustments. Furthermore, AI-driven rehabilitation guides the recovery process, tailoring routines to individual needs and optimizing return-to-play schedules. This chapter explores these cutting-edge technologies, demonstrating how AI not only mitigates injury risks but also enhances overall athlete well-being, laying a foundation for a healthier, more resilient future in field hockey.

Wearable Technology

Wearable technology has revolutionized the way athletes train, perform, and recover. In the realm of field hockey, these innovations are contributing significantly to injury prevention and management. The introduction of smart devices and sensors into wearable gear allows for real-time monitoring of an athlete's physical condition, providing invaluable data that can preempt injuries before they happen.

One of the main benefits of wearable technology is its ability to collect and analyze physiological data. For example, smartbands and fitness trackers equipped with heart rate monitors, GPS, and accelerometers offer a comprehensive look at an athlete's performance. By examining metrics like heart rate variability and exertion levels, coaches can identify signs of fatigue or overexertion that may precede injuries. This enables the team to implement rest and recovery protocols, tailoring training loads to each athlete's needs.

Moreover, advanced smart clothing embedded with sensors can track muscle activity and movement patterns. Compression garments, for instance, are integrated with electromyography (EMG) sensors that measure muscle activation. These insights are pivotal in identifying irregular movement patterns or muscle imbalances, which can be addressed through targeted strength and conditioning programs.

Injuries such as sprains and strains are common in field hockey, often resulting from repeated stress or improper technique. Wearable technology can mitigate the risk of such injuries by providing immediate feedback on biomechanics. For instance, smart insoles equipped with pressure sensors can analyze foot strike patterns and ground reaction forces. If an athlete's form deviates from the ideal, the system alerts them instantly, allowing for corrective measures.

The integration of wearable technology extends beyond just monitoring and prevention - it's also crucial for rehabilitation. Post-injury, smart wearables help in tracking the athlete's recovery progress. Devices that measure joint angles and ranges of motion can be used to ensure that a player's rehabilitation exercises are performed correctly, promoting efficient and safe recovery.

In addition, some wearables come with hydration sensors that measure sweat composition, helping to prevent dehydration-related issues such as cramps and heat exhaustion. By keeping athletes hydrated and aware of their fluid intake needs, the chances of dehydration-related injuries are minimized.

Beyond individual monitoring, team-wide analysis using aggregated data from multiple athletes can reveal trends and patterns that could signify underlying risk factors. Coaches and medical staff can utilize this data to implement preventive strategies, modifying training regimens and implementing rest periods for the team as necessary.

Importantly, wearable technology's impact isn't confined to training and games alone. It also extends to long-term athlete health. Continuous monitoring provides a historical perspective on an athlete's physical condition, helping to spot chronic issues early on. Teams can thus manage long-term health risks more effectively.

While wearable technology has already made strides in injury prevention and management in field hockey, the future holds even more promise. Emerging innovations like smart textiles, which are capable of monitoring a wide array of physiological signals, and advances in miniaturization are likely to make these technologies even more integrated into sports apparel and equipment. The ongoing development of AI-driven analytics will enhance these wearables, offering even more precise injury predictions and comprehensive health insights.

However, for all its benefits, the deployment of wearable technology also comes with challenges. Data privacy remains a critical issue - it's essential to ensure that athletes' personal information is securely handled. Moreover, the technology should complement, not replace, the expertise of coaches and medical professionals. Wearable tech should serve as an aid, providing additional layers of insight to enhance human judgment.

In summary, wearable technology is transforming the landscape of injury prevention and management in field hockey. By harnessing the power of real-time data and analytics, these devices offer a proactive approach to athlete health. As technology continues to advance, the synergy between AI and wearables promises even greater strides in keeping athletes safe, healthy, and performing at their best.

Predictive Injury Models

In the heightened competitive atmosphere of field hockey, predicting and preventing injuries can mean the difference between triumph and defeat. Predictive injury models are rapidly transforming how players, coaches, and medical staff approach injury prevention and management. Integrating AI into this domain offers a sophisticated approach to foreseeing injuries before they occur, enabling proactive intervention strategies that maintain peak performance.

Drawing on vast amounts of historical and real-time data, predictive models capitalize on machine learning algorithms to identify subtle patterns and risk factors that might not be immediately apparent to the human eye. These insights enable the creation of individualized injury profiles for each player, taking into account their unique biomechanical, physiological, and even psychological traits. Over time, these profiles become more accurate and nuanced, enhancing their predictive power.

One critical component of these models is the use of wearable technology. Advanced sensors embedded in wearables can monitor a myriad of parameters such as heart rate variability, muscle fatigue, joint load, and even sleep patterns. When analyzed collectively, this data can offer comprehensive insights into a player's condition, signaling early warning signs that precedent injury.

Consider Jane Doe, a forward who has consistently displayed impressive agility but has been plagued by recurring hamstring injuries.

61

Through a predictive injury model, her coaching and medical staff could recognize patterns in her movement data, indicating the need for targeted strength training and stretching routines to mitigate her risk. As a result, tailored interventions can be designed to not only improve her performance but also enhance her overall well-being.

However, the real magic happens when these predictive models are integrated with training regimens. Coaches can create dynamic training plans that adapt in response to the insights gleaned from these models. For instance, if a player shows signs of increased risk, their training intensity can be adjusted temporarily, balancing between pushing performance boundaries and ensuring safety.

Another game-changing aspect is real-time monitoring. On-field data gathering through wearables and other IoT (Internet of Things) devices means that any fluctuations in a player's physical condition can be quickly detected. Instant alerts can inform coaching and medical staff of immediate intervention needs, potentially averting serious injuries. Such dynamic, responsive systems ensure that decisions regarding a player's participation are data-driven rather than purely observational.

One might argue that these models hinge heavily on the quality and breadth of data. Diverse datasets contribute to the robustness of predictive models, improving their accuracy and reliability. Integrating varied data—ranging from physical assessments to even psychological questionnaires—enriches these models, ensuring that all potential risk factors are comprehensively addressed.

Moreover, the iterative nature of machine learning means that these models evolve. New data feeds continuously refine these algorithms, making them more adept at predicting injuries over time. This adaptive learning process ensures that as the sport evolves, so does the technology safeguarding its players.

Despite the promising benefits, there are inherent challenges. Data privacy remains a significant concern. Players must feel assured that their personal health data is protected and used exclusively for their benefit. Transparent communication about data usage and stringent security protocols are paramount in building and maintaining trust.

Additionally, there's the human factor to consider. Full reliance on predictive models may overlook individual player nuances that only experienced coaches or medical staff can identify. Thus, striking a balance between algorithmic recommendations and human judgment is crucial. Engagement with these models should be seen as complementary—enhancing, rather than replacing, the expertise of seasoned professionals.

Consider the case of implementing predictive models in a professional field hockey team: initially, players and staff might be skeptical. Through initial phases of observing and modest adjustments suggested by the models, the tangible benefits become evident. Reduced incidence of injuries and shorter recovery times gradually silence the skeptics, turning them into proponents of this technology.

Moving forward, combining predictive injury models with other AI-driven aspects of performance, like tactical and strategic intelligence, offers a holistic approach to player management. Integrating various facets of AI ensures a comprehensive system where each part reinforces the other, leading not only to injury prevention but also to optimized training and game performance.

In the end, the goal isn't just to prevent injuries but to foster an environment where athletes can consistently perform at their best. By embracing the advancements offered by predictive injury models, the entire ecosystem of field hockey—players, coaches, and medical staff—moves towards a future marked by enhanced safety and performance. The continuous evolution and application of AI in this domain promise to set new standards in how we understand and manage athlete health, leading to a new paradigm in sports readiness and excellence.

Rehabilitation and Recovery using AI

Injuries are an unfortunate but inevitable part of playing sports, and field hockey is no exception. The process of recovering from injuries has traditionally been slow and arduous, often requiring frequent visits to physical therapists, long hours of manual rehabilitation, and extensive monitoring. However, the advent of artificial intelligence (AI) is revolutionizing the way athletes approach both rehabilitation and recovery.

One of the most significant benefits AI brings to rehabilitation is personalization. Wearable technology equipped with AI algorithms can collect vast amounts of data on an athlete's movements, muscle activity, and overall physical condition. This data can then be analyzed to design highly specific and individualized rehabilitation programs. These programs adapt in real-time, making adjustments based on the athlete's progress, ensuring that they are always performing the most effective exercises for their condition. If an athlete is struggling with a particular drill, the AI system can immediately suggest modifications to reduce strain and speed up recovery.

Moreover, AI systems can monitor compliance with rehabilitation protocols. It's one thing to have an ideal plan, but ensuring an athlete adheres to it is another challenge. By using sensors and AI-driven applications, coaches and medical staff can track an athlete's adherence to their prescribed exercises and form. For instance, if an athlete needs to stretch a specific muscle group to prevent scar tissue build-up, the system can alert both the athlete and their therapist if the exercises aren't being performed correctly or frequently enough.

Aside from personalized plans, AI offers predictive insights that were previously unattainable. Through machine learning, AI can predict which athletes are at higher risk of re-injury by analyzing patterns in their movement and biomechanics. This method ensures that precautionary measures can be taken well in advance. For instance, if data shows that an athlete has a tendency to favor one leg due to a previous injury, strategies can be implemented to address this imbalance, reducing the likelihood of future injuries.

Moreover, one of the incredible advances in AI-driven rehabilitation is the development of virtual reality (VR) systems. VR provides immersive environments where athletes can perform rehabilitation exercises. These exercises can range from simple movements to simulations that mimic actual game conditions. For example, an injured field hockey player can practice their dribbling or shooting in a virtual setting that mirrors a real game. The beauty of VR is that it can be both engaging and effective, keeping athletes motivated while also helping refine their skills during recovery.

AI also aids in mental rehabilitation. Recovering from a serious injury isn't just a physical process; it's a mental one as well. Athletes often experience anxiety and a lack of confidence after an injury. AI can help here by using biofeedback techniques through wearables to monitor stress levels during rehabilitation exercises. If the system detects an increase in stress or anxiety, it can provide real-time feedback, such as suggesting breathing exercises or reminding the athlete to take a break. These mental health tools can be pivotal in ensuring that the psychological aspect of recovery is not neglected.

Furthermore, the integration of AI in rehabilitation isn't only beneficial for athletes; it also eases the burden on medical professionals. By automating the data collection and analysis processes, therapists can focus more on direct patient interaction and less on administrative tasks. This efficiency allows for better quality care and more time to ensure each athlete's unique needs are met.

The real-time feedback provided by AI systems extends to the realm of progress tracking. Athletes and their caregivers can monitor recovery through comprehensive dashboards that highlight improvements and areas that need more attention. These dashboards can be accessed on multiple devices, providing convenience and flexibility. For instance, a coach can check an athlete's progress from their phone while on the go, while an athlete can review their metrics from the comfort of home.

Consider the role of computer vision technology, another branch of AI, in rehabilitation. Advanced cameras and software can analyze an athlete's movements with great precision. This technology can detect subtle deviations in technique that might indicate a risk of exacerbation. Imagine an injured player performing a series of squats; computer vision can pinpoint a slight misalignment in knee position and suggest corrections, reducing the risk of further injury and enhancing the effectiveness of rehabilitation exercises.

Undeniably, AI's role extends to long-term recovery management. Some injuries, especially significant ones like ACL tears or severe fractures, require prolonged care even after the initial rehabilitation phase. AI can track these long-term progressions, adjusting recommendations as needed. If the data suggests a decline in performance or an increased risk of injury, immediate intervention steps can be taken to prevent deterioration.

The benefits of integrating AI into rehabilitation and recovery go beyond individual athletes. Teams and organizations can use aggregated data to identify common injury trends. Understanding these trends allows for the development of more effective training programs and injury prevention strategies that can be implemented across the board, creating a safer and more efficient training environment.

While the implementation of AI in rehabilitation is promising, it doesn't come without challenges. Ensuring data privacy and security is paramount, especially when dealing with sensitive medical information. Establishing ethical guidelines and transparent data-handling practices are crucial to gaining the trust of athletes and ensuring their data is used responsibly.

Moreover, accessibility to AI-driven rehabilitation should be a priority. Elite athletes and well-funded teams might have easy access to these tools, but the goal should be to democratize technology so that even amateur players or smaller clubs can reap the benefits. Partnerships with tech firms, subsidies, and open-source platforms might be ways to achieve this broader accessibility.

The use of AI in rehabilitation and recovery is a testament to how cutting-edge technology can dramatically improve athletes' quality of life and career longevity. As field hockey continues to evolve, one thing is clear: AI will play an increasingly vital role in keeping players healthy and at the top of their game. This exciting intersection of sports and technology promises a future where injuries are managed more effectively, recovery times are shortened, and athletes return to their fields stronger than ever before.

Chapter 7:
Recruiting and Talent Identification

In the realm of field hockey, recruiting the next generation of talent is a game-changer, and artificial intelligence is primed to revolutionize this critical process. Forget the days of scouts tirelessly traveling from match to match; now, AI-driven solutions offer unparalleled precision and efficiency in talent identification. By leveraging advanced algorithms and machine learning, we can analyze a myriad of metrics—ranging from physical attributes and skill levels to psychological traits and performance under pressure. This holistic approach ensures that no promising athlete flies under the radar. Moreover, AI can predict an athlete's growth trajectory, making it easier for coaches and scouts to make informed decisions. Integrating these tools into recruitment strategies does more than just streamline processes—it democratizes opportunities, ensuring that talent from diverse backgrounds is recognized and nurtured. The future of field hockey is bright, as technology and human expertise work hand in hand to uncover and develop the stars of tomorrow.

Scouting Tools and Techniques

In the evolving landscape of field hockey, scouting has become more sophisticated and data-driven, thanks to advanced tools and techniques. No longer reliant solely on the keen eyes of seasoned scouts, modern talent identification leverages a variety of technological aids to dig deeper into a player's potential and performance.

One of the most transformative tools in scouting is video analysis. This isn't just about watching recorded games anymore. High-definition cameras and specialized software enable scouts to break down every aspect of a player's game. By analyzing hundreds of hours of footage, scouts can identify nuanced patterns and tendencies that might otherwise go unnoticed. This granular level of detail is invaluable for assessing a player's decision-making, positioning, and technical skills under different conditions.

Additionally, wearable technology has revolutionized the scouting process. Devices such as GPS trackers and heart rate monitors provide real-time data on a player's physical performance, including speed, endurance, and recovery time. By analyzing this data, scouts can gauge a player's athleticism and overall fitness, offering a more rounded view than what is visible during gameplay.

Machine learning algorithms play a crucial role in augmenting traditional scouting methods. By feeding these algorithms a vast amount of performance data, they can identify statistical patterns and correlations that might not be immediately obvious. This predictive modeling helps pinpoint players who not only excel currently but have the potential to develop into top-tier talent. Essentially, it's like having a second set of eyes that never tire and can scrutinize data at a scale that human scouts simply can't match.

Another innovative approach is the use of AI-powered data analytics. Advanced algorithms evaluate diverse datasets—ranging from social media metrics to academic records—to provide a holistic profile of each player. Such comprehensive insights help in understanding a player's mental fortitude, leadership qualities, and how they might fit into team dynamics. It's not just about finding the most skillful player, but finding the right player for the team's culture and long-term goals.

The rise of virtual reality (VR) is also making waves in scouting. VR simulations can recreate in-game scenarios, allowing players to showcase their abilities in a controlled environment. For scouts, this means observing how a player reacts to specific situations without the need for live practice sessions, which may be logistically challenging. This tool is particularly useful when assessing younger talent who might not yet have extensive game footage available.

Beyond individual tools, the integration of multiple technologies into a cohesive scouting system represents the cutting edge of talent identification. For instance, merging video analysis with wearable tech data and machine learning insights creates a multidimensional profile of a player. This holistic approach enables scouts and coaches to make data-informed decisions, significantly reducing the risks associated with talent recruitment.

However, the best results come from combining these advanced tools with the traditional wisdom and intuitive insights of experienced scouts. Technology can point out a player's strengths and weaknesses through quantitative data, but the qualitative aspect of scouting—such as understanding a player's passion, work ethic, and adaptability—still requires a human touch. The integration of human expertise and technological innovation ensures a well-rounded scouting process that maximizes the chances of finding and nurturing future stars.

Moreover, the data gathered from these tools doesn't just stay with the scouts. It often gets shared with coaches and players, forming the basis for personalized training programs. For aspiring players, this feedback loop is invaluable. It allows them to identify areas for improvement, track their progress, and align their development with the expectations of elite selectors. In essence, scouting tools and techniques don't just identify talent—they actively contribute to the nurturing of that talent.

Another emerging trend in scouting is the use of drone technology. Drones provide aerial views of games and practice sessions, offering unique perspectives that ground-level cameras can't achieve. These aerial views can reveal a team's structural formations and a player's spatial awareness and movement patterns within the team's setup. This bird's-eye view can be particularly insightful for analyzing defensive strategies and transitions, which are critical in high-stakes matches.

Social media analytics is another unconventional yet powerful scouting tool. Players often display their personalities and share their career milestones online. By analyzing social media activity, scouts can gain insights into a player's character, resilience, and public presence. These digital footprints can influence decisions, especially when gauging a player's potential to handle the spotlight and their fit within the societal values of the team or organization.

Furthermore, artificial intelligence aids in creating comprehensive player databases that can be accessed and updated in real-time. Scouting teams can keep track of numerous prospects simultaneously, efficiently comparing stats and performances across different leagues and competitions. These databases also facilitate quick cross-referencing and trend identification, helping scouts uncover hidden gems who might otherwise remain under the radar.

For international scouting, language barriers and regional scouting biases often pose challenges. Here, AI-powered translation tools and unbiased data analytics level the playing field, ensuring that talent is recognized regardless of geographical location or language proficiency. This global approach to scouting broadens the talent pool and ensures that no stone is left unturned in the quest for excellence.

As technology continues to evolve, so too will the tools and techniques used for scouting. The constant flow of new data and the ever-improving capabilities of AI systems mean that scouting will become even more precise and predictive. While this might sound like science fiction, it's very much the reality of the modern field hockey environment. And as these tools become more refined, they will only serve to bring out the best in the players they identify, ensuring that the sport continues to reach new heights.

The future of scouting in field hockey is both exciting and promising. The integration of advanced tools ensures that talent identification is more comprehensive, efficient, and fair. This technological evolution not only aids scouts and coaches in making informed decisions but also empowers players to achieve their fullest potential, thereby elevating the entire sport.

AI in Talent Assessment

Artificial intelligence is revolutionizing how field hockey teams scout and evaluate talent, making the process faster, more efficient, and potentially less biased. This transformation is critical in a sport where the difference between winning and losing often comes down to the slight edge conferred by spotting and nurturing exceptional talent early. AI-based talent assessment thus steps in as a game-changer, equipping coaches and scouts with tools that were unimaginable a decade ago.

Conventionally, talent assessment in field hockey, much like in other sports, relied heavily on subjective judgment. Coaches and scouts traveled far and wide, watching countless matches and tryouts, often relying on gut feelings and instinct. AI's introduction adds a layer of objectivity, offering data-driven insights that can quantify attributes such as speed, decision-making ability, and even psychological resilience. Imagine an algorithm that not only identifies the fastest sprinter on the field but also predicts who can maintain peak performance under high-pressure scenarios. This evolution is radical and potentially transformative.

AI excels in pattern recognition—one of its most potent features when applied to talent assessment. By analyzing vast troves of data from player performance—think GPS trackers, video analysis, and even social media behavior—AI can identify patterns and outliers that human scouts might overlook. For instance, machine learning algorithms can process video footage frame by frame to analyze a player's reaction times, passing accuracy, or spatial awareness. Such granular analysis helps identify players who may not be the stars in obvious metrics like goal-scoring but who make those around them better, the hidden gems.

Video analysis plays an integral role in AI-powered talent assessment. With advancements in computer vision, AI can now break down game footage into minute actions and movements. A simple pass or a defensive maneuver can be tagged, measured, and compared against thousands of other instances. Automated systems can identify strengths and weaknesses that are not immediately apparent to the naked eye and provide detailed reports. These insights allow coaches to make more informed decisions about which players to focus their development efforts on.

A significant advantage of using AI in talent assessment is its ability to analyze not just current performance, but also potential for future development. Machine learning models can ingest data over time, tracking a player's progress and adaptations. They can predict future performance trajectories and even suggest tailored training programs to maximize a player's growth. For young athletes, particularly, this kind of targeted development can be a game-changer, helping them capitalize on their potential earlier and more effectively.

Moreover, AI isn't limited to analyzing physical performance. Sophisticated algorithms can scrutinize psychological attributes, too. By analyzing data collected from psychometric tests, social media activity, and even interviews, AI can provide a comprehensive profile of a player's mental makeup. Attributes like grit, teamwork, and focus—all crucial elements in high-stress game situations—can be gauged with surprising accuracy. This holistic approach ensures that teams aren't just choosing the most physically gifted players but those who exhibit the psychological traits essential for success.

Bias in talent identification is a persistent challenge in sports. Whether conscious or unconscious, human scouts may favor players who conform to certain preconceived ideals, thus potentially overlooking diverse talent. AI can mitigate this issue by focusing strictly on data. Gender, ethnicity, or socioeconomic background become irrelevant in front of models that prioritize performance metrics. This democratization of talent assessment ensures a fairer and more inclusive approach, identifying players based purely on their skills and potential.

The ability to access and analyze data from around the globe opens up new avenues for talent discovery. Teams no longer need to limit their searches to local or national pools; they can scout for talent internationally with unprecedented ease. Sophisticated AI platforms aggregate data from various leagues and tournaments worldwide, providing a more comprehensive talent pool. Whether it's tracking promising young players in European leagues or uncovering hidden stars in lesser-known tournaments in Asia or Africa, the scope for discovery increases exponentially.

Integration of AI in scouting also brings about a more streamlined collaboration between different stakeholders in a team. Coaches, analysts, and medical staff can all tap into a shared database of player assessments. A unified data platform ensures that everyone is on the same page, making the talent identification process more seamless and less fragmented. This collaborative approach enables a more holistic evaluation of players, considering inputs from various experts who bring different perspectives and insights to the table.

By reducing the time and resources spent on the initial stages of talent identification, AI allows human experts to focus more on the nuances that machines can't yet fully grasp. For instance, while AI can offer detailed statistics and predictions, the gut instinct of a seasoned coach still plays a vital role in final decisions. AI tools don't replace human judgment but rather augment it, providing a robust framework within which human expertise can operate more effectively.

AI also plays a crucial role in the continuous improvement of scouting models. Machine learning algorithms evolve with new data, adapting and refining their predictions over time. As more and varied data is fed into the system, the models become increasingly sophisticated, capable of making better, more accurate assessments. Teams that invest in these technologies often find themselves at the cutting edge, benefiting from a self-improving system that keeps them ahead of the curve.

Another emerging aspect is the ethical considerations surrounding AI usage in talent assessment. The algorithms are only as good as the data fed into them, and biases can creep in if the data set isn't appropriately managed. Transparency in how algorithms make decisions is crucial to maintain trust among all stakeholders. Teams must be vigilant about continually auditing and refining their AI systems to ensure they meet ethical standards and deliver on their promise of fair, unbiased talent assessment.

In conclusion, AI's growing role in talent assessment is reshaping how field hockey teams scout, evaluate, and nurture talent. By leveraging data-driven insights and sophisticated algorithms, teams can make more informed, objective, and inclusive decisions. The tools at their disposal are becoming more advanced and nuanced, offering a granular analysis that was previously unattainable. While AI doesn't replace the human touch, it significantly augments it, providing a potent combination that could well define the future of talent identification in field hockey.

Developing Future Stars

At the core of any successful field hockey program is the ability to identify and nurture talent. "Developing Future Stars" takes this concept and amplifies it with the power of artificial intelligence. Instead of relying solely on traditional methods, coaches and scouts can now use AI to pinpoint burgeoning talent, making the process more efficient and effective.

The competitive nature of field hockey means that spotting talent early can be the difference between good and great teams. AI systems analyze vast amounts of data that human eyes might overlook. By studying player statistics, performance metrics, and even physiological data, these systems create profiles that highlight potential stars who might otherwise go unnoticed. This isn't just about recognizing a player's current skill level; it's about predicting their trajectory and long-term potential.

One of the key ways AI aids in talent development is by creating personalized development plans. These plans cater to an individual's strengths and weaknesses, optimizing training sessions to fast-track improvement. Instead of one-size-fits-all training regimes, players get tailored workouts and drills. This approach not only accelerates learning but also helps in preventing injuries by avoiding overuse and fostering balanced physical development.

The influence of AI extends to tracking a player's mental attributes, like decision-making speed and resilience. Machine learning algorithms can analyze gameplay footage to understand how a player reacts under pressure, their tactical awareness, and their adaptability on the field. This allows coaches to focus not just on physical training but also on mental toughness and strategic thinking—crucial aspects that differentiate average players from elites.

Moreover, AI-driven tools are transforming traditional scouting techniques. Scouts armed with AI apps or systems can input live data during matches, generating immediate insights about a player's performance. These tools can even compare players against established benchmarks or past performances to identify who consistently excels or shows promise. This level of detail and immediacy was previously unattainable.

It's not just about data collection, though. The way AI processes this data leads to actionable insights. When scouts and coaches gather to discuss player potentials, they have concrete, data-driven evidence to guide their decisions. This minimizes biases and subjective opinions, ensuring the selection process is as fair and accurate as possible.

But the role of AI doesn't stop at identifying young talent. It continues to support them through their development journey. Wearable technology, for instance, can monitor players' workloads, physical conditions, and recovery times, providing real-time feedback and adjustments to training sessions. If a young player shows signs of fatigue, the system will recommend lighter sessions. If they're on the verge of a breakthrough in a particular skill, the AI might suggest more focused drills.

Wearable sensors monitoring player health

Predictive analytics flagging potential injuries

Machine learning algorithms optimizing training loads

AI tools evaluating mental attributes

Data-driven guidance for talent development

Communication is another pivotal aspect where AI lends a hand. With the use of digital platforms powered by AI, young athletes can receive instant feedback. Instead of waiting for post-match reviews, they get on-the-spot coaching tips. This immediacy helps in making quick corrections, fostering a quicker learning curve. Through video analysis, players can see their movements, compare them to ideal scenarios, and understand abstract concepts visually and intuitively.

Artificial intelligence also acts as a bridge between different levels of the sport. From grassroots to elite academies, AI systems ensure consistent tracking and development of players as they progress through the ranks. This continuity is key to maintaining a coherent developmental pathway. When a player transitions from a local club to a national team, their entire developmental history is available, allowing for a seamless integration into more advanced training programs.

For all its benefits, the human element remains irreplaceable. AI provides the tools for better analysis, training, and decision-making, but it is the coaches, mentors, and players themselves who bring the game to life. AI's role is to support and enhance human judgment, not to replace it. The rapport between player and coach, the intuition of a seasoned scout, and the player's passion and drive—these are elements that technology will never replicate but will always aim to amplify.

Looking ahead, the integration of AI in developing future stars will only grow stronger. As technology evolves, the tools at our disposal will become even more sophisticated. Imagine a future where AI not only helps in identifying raw talent but also predicts the best possible career paths for players. By analyzing market trends, game evolutions, and individual capabilities, AI could provide tailored career advice that aligns players' skills with the sport's future demands.

The vision for developing future stars is a holistic one. It's about creating an environment where players can thrive, maximizing their potential through the intelligent use of technology. It's about setting the stage for the next generation of field hockey legends, ensuring the sport remains dynamic and exciting. With AI as an ally, the path from novice to star becomes clearer, more efficient, and incredibly promising.

The future of field hockey is brimming with potential, and those who harness the power of AI in developing talent will undoubtedly be at the forefront of the game. By blending the best of technology with the irreplaceable human elements of passion, intuition, and mentorship, we can pave the way for the stars of tomorrow in a way that was previously unimaginable. The next chapter awaits, filled with opportunities to uncover and cultivate the field hockey greats of the future.

Chapter 8:
Fan Engagement and Experience

Field hockey is not just a game played between the lines; it is an experience that enthralls every fan with a balance of tradition and modernity. In this digital age, artificial intelligence is drastically transforming how fans engage with the sport. AI-powered broadcasting and commentary are creating tailored viewing experiences, allowing fans to delve deeper into the strategies and nuances of every match. Enhanced interaction through interactive technologies makes fans feel more connected to their favorite teams and players. Virtual and augmented reality are breaking boundaries, offering immersive experiences that bring the stadium right into your living room. These advancements are revolutionizing fan engagement, making it more dynamic and personalized, thus ensuring that the passion for field hockey not only stays alive but thrives in the era of technology.

AI in Broadcasting and Commentary

In the continually evolving landscape of field hockey, artificial intelligence has found a unique and impactful role in broadcasting and commentary. This transformation has not only redefined how games are viewed but also significantly enhanced fan engagement. Advances in AI are providing broadcasters with smart tools to deliver in-depth analysis, real-time insights, and enriched storytelling, making the viewer experience more immersive and informative than ever before.

Artificial intelligence enhances the traditional role of commentators by providing them with real-time data and predictive insights. For instance, AI algorithms can analyze player movements, game statistics, and historical data to offer predictions about potential game outcomes. This data is then seamlessly integrated into the broadcast, offering viewers a level of insight that was previously only accessible to professional analysts. Imagine watching a game and, as a player sets up for a penalty corner, the broadcast shows the likelihood of scoring based on variables such as the player's past performance, the angle of the shot, and the goalkeeper's save history. Such in-depth analysis was unimaginable just a few years ago.

Moreover, AI-powered commentary systems can adapt to different broadcasting styles and regional preferences. They can generate multilingual commentaries, enhancing accessibility and engagement for a global audience. By employing natural language processing and machine learning, these systems can understand and replicate the nuances of human speech. Therefore, international fans can enjoy personalized commentaries in their native languages without missing the excitement and critical insights delivered during the game.

Machine vision, another facet of AI in broadcasting, plays a pivotal role in enriching the viewer's experience. By tracking player movements and analyzing game patterns, machine vision tools can generate heat maps, player positioning graphics, and dynamic replays that highlight crucial moments with unprecedented clarity. These visual aids can be overlaid on the live broadcast, helping fans better understand game dynamics and player strategies. The integration of such technology not only deepens engagement but also educates viewers, turning casual spectators into more knowledgeable fans.

Additionally, AI-driven sentiment analysis tools are employed to gauge fan reactions in real-time through social media and other platforms. This data provides broadcasters with valuable insights into viewer preferences and trends. By understanding what excites or frustrates fans, broadcasters can tailor their content to keep the audience captivated. For example, if a significant portion of the audience is thrilled about a particular player's performance, commentators can focus more on that player's contributions, thereby enhancing viewer satisfaction.

Integrating AI into broadcasting also opens the door for interactive and personalized viewing experiences. Platforms now offer features where fans can customize their viewing stats, select different camera angles, and even receive notifications about key moments they might have missed during the live game. This level of interactivity ensures that fans remain engaged, even if they're unable to watch the entire match live. Such features turn passive viewers into active participants in the game.

Virtual and augmented reality tools, empowered by AI, are revolutionizing the way games are consumed. Augmented reality can bring additional layers of information to the screen, such as player stats hovering over their heads as they move, or a virtual scoreboard with real-time updates. Virtual reality can transport fans directly to the field from the comfort of their homes, offering a 360-degree view of the game. These experiences are enriched with AI-generated real-time data, creating an unparalleled sense of immersion and allowing fans to experience the thrill of the game from entirely new perspectives.

The use of AI in broadcasting also brings in an element of democratization. Historically, high-quality game analysis and commentary were restricted to major games and leagues with substantial budgets. However, AI tools are becoming increasingly accessible, enabling lower-tier leagues and amateur games to deliver a professional-grade viewing experience. This democratization not only boosts the popularity of field hockey at grassroots levels but also helps in discovering and nurturing new talent by bringing their performances to a broader audience.

As AI continues to evolve, its applications in broadcasting and commentary will undoubtedly become more sophisticated. Predictive analytics will get sharper, machine learning models will become more accurate, and natural language generation will be more human-like. As a result, fans will enjoy more engaging, informative, and personalized viewing experiences. This transformation signifies a new era in sports broadcasting, where technology and human expertise join hands to offer a comprehensive and exhilarating experience.

Looking forward, the integration of AI in broadcasting and commentary will not only enhance fan engagement but also shape the future of sports viewing. Broadcasters will leverage AI to create more compelling narratives, young fans will grow up interacting with these advanced systems, and a deeper connection between the sport and its followers will be forged. AI isn't just a tool for the present; it's an investment in the future of field hockey, promising a viewing experience that's rich, interactive, and smarter. AI in broadcasting and commentary is not just about changing the game; it's about harnessing technology to celebrate the sport in ways we could have only dreamed of.

Enhanced Fan Interaction through Technology

In today's digital world, the line between spectators and participants is increasingly blurred. Nowhere is this more evident than in sports, particularly field hockey. For fans, the thrill of the game used to be limited to just watching from the stands or on television. With advances in technology, especially artificial intelligence, fan interaction has become more immersive and engaging than ever before. AI-driven platforms are transforming how fans experience the game, offering both broad insights and minute details that were once accessible only to players and coaches.

Consider the role of social media in reshaping fan interaction. Social platforms powered by AI algorithms curate content that caters to individual preferences, making it easy for fans to engage with their favorite teams, players, and moments in real-time. More than just a one-way stream of updates, fans can now participate in live polls, ask questions during broadcasts, and even make real-time predictions, actively influencing the narrative around the game.

AI is also revolutionizing how fans get their news and updates. Automated systems can generate match summaries, player statistics, and predictive analyses. Journalists and commentators benefit from these tools, which help them produce richer, more engaging content. Consequently, fans receive a more nuanced understanding of the game, enhancing their overall experience and engagement.

Moreover, AI-integrated mobile apps offer fans a more personalized experience. Through these apps, supporters can receive tailored notifications about player stats, game results, and live score updates. Imagine an app that not only provides the scores but also breaks down complex plays, offers predictive game outcomes, and even provides historical context—all in the palm of your hand.

Artificial Intelligence isn't just confined to passive updates and notifications. Interactive features like chatbots are becoming increasingly popular. These AI-driven bots simulate conversation and provide instant responses to fan queries, ranging from game schedules to player trivia. They can even offer personalized content based on user interactions and preferences. The immediacy and personalization offered by these tools make fans feel connected to the sport in ways they never have before.

Another exciting development is the use of augmented reality (AR) and virtual reality (VR) to elevate the fan experience. Fans unable to attend live games can enjoy the next best thing—immersive experiences that make them feel like they're right in the stadium. With AR and VR, fans can virtually walk through the team's locker room, watch the game from any angle, and even 'meet' their favorite players. These technologies bridge the gap between physical and digital experiences, making field hockey more accessible to a global audience.

Field hockey clubs and organizations are also exploring how AI can enhance fan engagement through loyalty programs. Imagine using machine learning algorithms to analyze fan behavior and tailor rewards accordingly. By understanding what excites and motivates their fanbase, clubs can offer customized experiences that increase engagement and loyalty. From exclusive content and behind-the-scenes footage to personalized merchandise recommendations, the possibilities are endless.

Live broadcasting has also become more sophisticated thanks to AI. Modern broadcasting involves real-time data overlays, where fans get instant access to detailed statistics, player performance metrics, and even predictive analytics during the game. For example, AI-powered cameras can automatically track the action on the field, providing uninterrupted, high-quality coverage. A fan's viewing experience is enriched by a flood of data and insights, making each game both informative and exciting.

Speaking of broadcasting, one can't overlook the advancements in commentary made possible by AI. Intelligent systems can now provide play-by-play analysis, identify key moments, and even predict future plays. These enhancements make broadcasts more dynamic and capture the attention of both casual viewers and die-hard fans. It's as if every fan has a personal commentator guiding them through the intricacies of the game.

Additionally, AI helps in creating unique fan experiences during actual events. Smart stadiums equipped with IoT (Internet of Things) devices can integrate AI to offer a seamless and interactive experience. Imagine entering a stadium where your ticket is authenticated through facial recognition, where an app guides you to your seat, and where you can order snacks and merchandise without leaving your spot. Such advancements turn attending a game into an interactive adventure, making fans feel more integral to the event.

Gamification is another area where AI is making strides. Loyalty programs, fantasy leagues, and trivia competitions are being optimized through machine learning to increase engagement. By analyzing past behavior, these AI systems can fine-tune the difficulty levels, offer relevant rewards, and keep fans coming back for more. This added layer of interaction makes fans feel like part of the team, heightening emotional investment in the sport.

The role of AI in fan engagement extends beyond just the individual experience. It also allows for community building by facilitating better fan-to-fan interactions. Online forums, social media groups, and team-based challenges are optimized through algorithms that recommend content, connect like-minded fans, and foster a sense of community. When fans feel connected not only to their teams but also to each other, the overall fan atmosphere becomes more vibrant and engaging.

Yet, with all these advancements, it's crucial to strike a balance. While AI can automate and enhance many aspects of fan engagement, the human touch remains irreplaceable. The unquantifiable excitement, the human emotions, and the unpredictable nature of sports are what make field hockey captivating. Therefore, the integration of AI should aim to complement the human elements rather than overshadow them.

To sum up, the impact of AI on fan engagement in field hockey is profound and transformative. From personalized content and interactive experiences to enhanced broadcasting and loyalty programs, technology is bringing fans closer to the action in ways previously unimaginable. As AI continues to evolve, one can expect even more innovative solutions that will further dissolve the boundaries between the fans and the game they love.

Virtual and Augmented Reality

In the ever-evolving landscape of sports, the integration of virtual and augmented reality (VR/AR) is rapidly reshaping the way fans engage with field hockey. Far beyond just flashy tech add-ons, these immersive technologies have the potential to amplify the fan experience, making matches more exhilarating and interactive than ever before. From live game broadcasts to fan participation in training simulations, VR and AR are set to revolutionize how spectators interact with the game.

Imagine donning a VR headset and finding yourself in the center of a bustling field hockey stadium, miles away from your actual location. Every cheer, every skirmish, and every breathtaking goal experienced as if you were physically there. The fluidity and immediacy VR brings to live game streaming offers fans a front-row seat, right from the comfort of their homes. By eliminating logistical barriers, VR ensures that the thrill of a live match is accessible to every fan, regardless of their geographic location.

Beyond live streaming, augmented reality is transforming the game-day experience inside the stadium. AR applications can superimpose player statistics, historical data, and real-time analytics over the live view, accessible through smartphones or AR glasses. Picture this: You're watching a penalty corner unfold and, with the flick of a wrist, up pops details about the shooter's past performance in similar situations. This amalgamation of real-world action with digital information creates a richer, more informed viewing experience.

For those who love dissecting game strategies, AR brings a whole new dimension. Platforms are emerging that allow fans to step into a coach's shoes, assessing player positions and suggesting tactics in real-time during the game. This interactive layer of engagement helps fans understand the complexities and nuances of the sport, fostering a deeper connection to the game. Games are no longer just watched; they are analyzed, critiqued, and even co-strategized by the audience.

Training simulations using VR are another groundbreaking aspect. While primarily designed for player development, these simulations can offer a unique, participatory experience for fans as well. Interactive sessions where fans can "train" alongside their favorite players or take penalty shots against a virtual goalie have a dual impact: they educate while entertaining. Fans gain a newfound appreciation for the skill and athleticism required in field hockey, all while feeling actively involved.

Incorporating VR and AR into fan engagement also paves the way for innovative fan experiences, such as virtual meet-and-greets with players or behind-the-scenes tours of locker rooms and training facilities. These exclusive experiences, once limited to a select few, become democratized through technology, building loyalty and enhancing the emotional bond fans have with their teams.

Additionally, the integration of VR and AR extends beyond game days and into fan communities. Social VR platforms can create virtual stadiums where fans from around the world gather to watch matches together, complete with cheering, discussions, and live reactions. These virtual communities foster a sense of belonging and shared passion, making the fan experience more social and immersive.

This transformation isn't just limited to the here and now; the future possibilities are expansive. Imagine interactive holographic replays where fans can walk around the digital renderings of critical game moments, examining plays from every conceivable angle. Such advancements would offer deeper insights and a more immersive analysis than ever before, making the viewer part of the action.

Integration of VR and AR technology does come with its challenges. Ensuring seamless and lag-free experiences requires robust infrastructure and consistent technological advancements. However, the potential benefits far outweigh these hurdles. Sponsorships and ad revenues could see a significant boost as companies rush to be part of these cutting-edge engagement tools, providing fresh revenue streams for field hockey organizations.

Consequently, as VR and AR technologies continue to evolve and mature, they hold the promise of transforming passive spectators into active participants. The revolution in fan engagement isn't just about watching; it's about feeling, understanding, and participating in the sport on a whole new level. The visceral connection forged through these immersive technologies could well be the future of field hockey, creating an inclusive, engaging environment where technology bridges the gap between the game and its ardent followers.

The horizon is immensely promising, and the field hockey community stands to gain immensely from embracing these innovations. As more teams and organizations adopt VR and AR, the line between the real and the virtual blurs, offering fans an unparalleled, immersive experience. These advancements herald a new era where technology doesn't overshadow the sport but enhances it, weaving deeper connections between the game and its global fanbase.

In conclusion, VR and AR offer brilliant, multifaceted ways to elevate field hockey's appeal and accessibility. They provide fans with immersive experiences that deepen their engagement and understanding of the game, turning casual watchers into die-hard enthusiasts. As we move forward, embracing these technologies will be pivotal in keeping the sport vibrant, exciting, and deeply connected with its audience.

Chapter 9:
Ethical Considerations of
AI in Field Hockey

As AI continues to transform field hockey, ethical considerations become paramount. Ensuring data privacy and security for players and teams is crucial, given the sensitive nature of performance metrics and personal health information. Beyond this, the fairness and bias in AI approaches need addressing to avoid skewed results that might advantage or disadvantage certain players or teams. Balancing AI and human judgment remains essential to preserve the human touch intrinsic to coaching and decision-making. While AI offers unprecedented insights and efficiencies, the integrity of the sport must be upheld, ensuring technology complements rather than overshadows the human spirit and skill that define field hockey.

Data Privacy and Security

Ensuring data privacy and security is crucial in the intersection of AI and field hockey. With the proliferation of AI technologies in various aspects of the sport, from training and performance analysis to fan engagement, massive amounts of data are generated and processed. This data often includes personal information about players, such as biometric data, performance metrics, and even behavioral patterns. Protecting this information is not just a matter of regulatory compliance; it's about maintaining the trust of athletes, coaches, and other stakeholders.

The first concern revolves around the collection of data. AI systems used in field hockey gather extensive data from various sources such as smart wearables, video analytics, and tracking devices. Each of these sources can capture sensitive information, requiring robust measures to secure it from unauthorized access. It is crucial to implement encryption protocols and access controls to ensure that only authorized personnel can view or interact with the data. Simple password protections are no longer sufficient; multi-factor authentication and regularly updated security protocols are essential.

Moreover, it is important to consider the anonymization of data. By anonymizing personal data, the risk of compromising individual privacy is significantly reduced. This is particularly critical when the data is used for research or published in any form. Anonymized data allows for valuable insights without exposing individual identities, striking a balance between innovation and privacy.

Another layer of complexity is added by data sharing. In the interconnected ecosystem of field hockey, data often needs to be shared across different platforms and stakeholders including coaches, analysts, and even third-party tech providers. Each point of transfer is a potential vulnerability. Secure APIs and encrypted communication channels should be employed to safeguard data during transfer. Furthermore, implementing meticulous logging and monitoring helps in detecting and responding to any unauthorized data access in real-time.

Apart from external threats, internal threats also pose significant risks. Employees or insiders with access to sensitive data could misuse it for malicious purposes or personal gain. It is essential to establish stringent internal policies and regular audits to mitigate these risks. Training programs that focus on data handling practices and ethical considerations should be mandatory for staff members who interact with sensitive data. These programs can raise awareness about the importance of data privacy and security, thereby fostering a culture of responsibility.

Compliance with legal frameworks is another critical aspect. Various regulations such as the General Data Protection Regulation (GDPR) and the California Consumer Privacy Act (CCPA) set stringent guidelines for data protection. Ensuring compliance with these regulations is not just a legal requirement but also a way to fortify trust among stakeholders. Regular audits and compliance checks can identify potential gaps in data protection practices, allowing for timely remediation.

Beyond compliance, transparency plays a pivotal role in data privacy and security. Organizations and teams should be transparent about their data collection practices and the measures they take to safeguard it. Privacy policies must be clear and accessible, providing stakeholders with a comprehensive understanding of how their data is handled. Options for individuals to control their data, such as consent forms and opt-out mechanisms, should be readily available and easy to use.

The ethical implications of data privacy extend to bias and fairness in AI systems. Biometric and performance data used to train AI models must be representative of all athletes to avoid biased outcomes. If data privacy measures inadvertently exclude certain groups, the resulting AI models could perpetuate biases, leading to unfair advantages or disadvantages. Ensuring inclusivity in data collection practices is paramount to the integrity and fairness of AI applications in field hockey.

While preventive measures are vital, preparedness for potential data breaches is equally important. Developing a robust incident response plan that outlines the steps to be taken in the event of a data breach can mitigate the impact on individuals and the organization. This plan should include immediate actions such as containing the breach, notifying affected parties, and cooperating with regulatory bodies. Conducting regular drills and simulations can help in refining the incident response plan and ensuring its effectiveness.

In today's digital age, the cybersecurity landscape is continually evolving. New vulnerabilities and threats emerge, necessitating constant vigilance and adaptability. Leveraging artificial intelligence for security purposes can offer advanced threat detection mechanisms, pinpointing anomalies that human oversight might miss. AI-driven cybersecurity tools can proactively identify and neutralize threats, providing an extra layer of protection for sensitive data.

It is also worth noting the role of collaboration in enhancing data privacy and security. Field hockey organizations and stakeholders can benefit from sharing knowledge and best practices related to cybersecurity. Industry forums, workshops, and collaborative projects can foster a collective effort to tackle common challenges, thereby raising the standard of data protection across the board.

To summarize, safeguarding data privacy and security is an indispensable aspect of integrating AI into field hockey. From collecting and anonymizing data to ensuring compliance and preparing for breaches, each step requires meticulous attention. It's a proactive, ongoing process that demands a combination of technological measures, ethical considerations, and collaborative efforts. Ensuring robust data privacy and security not only complies with legal standards but also builds trust and fosters a healthier, more equitable environment for all stakeholders involved in the sport.

Fairness and Bias in AI Approaches

As field hockey increasingly embraces artificial intelligence (AI) in its training, performance analysis, and game strategy development, the issue of fairness and bias within AI systems becomes paramount. The field of AI is not immune to biases; in fact, the algorithms and data that drive these technologies can inadvertently amplify existing biases or introduce new ones. It's essential to recognize these risks and implement strategies to ensure AI applications contribute to a fair and level playing field.

Bias in AI can stem from several sources. Primarily, it arises from the data used to train AI models. If historical data contains biases—whether relating to gender, race, or socioeconomic status—the AI systems can learn and perpetuate these biases. For example, if an AI talent scouting tool primarily trained on data from male players, it may be less effective—or outright biased—when assessing female athletes. This can severely impact the fairness in recruitment processes, potentially overlooking talented players due to biased algorithms.

Ensuring fairness starts with the data collection process. Data used to train AI models must be comprehensive, representing diverse demographics and varying levels of play. This inclusivity helps mitigate biases and allows the AI to make equitable assessments and recommendations. Additionally, regular audits of AI systems can detect and address bias early, ensuring these technologies evolve to serve all players and teams fairly.

Another critical area where bias can manifest is in AI-driven coaching and training tools. Personalized training programs powered by AI can offer immense benefits, but they must be carefully designed to avoid fairness pitfalls. If AI models are not calibrated to account for different playing styles, physiological differences, or individual player development paths, they risk promoting a one-size-fits-all approach. This can disadvantage players who may not conform to the 'norm' captured in the model's training data but possess unique talents and potential.

To combat this, developers and coaches must work collaboratively to tailor AI tools to the needs of a diverse player base. Regular feedback loops between players, coaches, and AI developers can ensure that the tools remain adaptive and fair. Moreover, incorporating human oversight in AI-driven coaching can balance the objectivity of AI with the nuanced understanding coaches bring to the game.

In performance analysis, biases can also creep in if the AI models prioritize certain metrics over others. For instance, if an AI system overemphasizes physical metrics such as speed or strength and underestimates cognitive aspects like strategic thinking, it may unfairly advantage physically dominant players. This could skew talent identification and team selection processes, potentially sidelining players who excel in strategic and mental dimensions of the game.

To create a fair evaluation landscape, it's vital to develop multi-faceted AI models that consider a variety of performance metrics. These should balance physical attributes with cognitive and strategic skills, providing a more holistic assessment of a player's capabilities. Ensuring transparency in how these metrics are weighted and used by AI systems can also help players and coaches trust the technologies and work with confidence.

Bias can also affect real-time game strategy recommendations generated by AI. For example, if an AI model has mainly analyzed data from a particular style of play or league, its recommendations might be tailored to that specific context, potentially misaligning with the team's unique strategy or the current game dynamics. This can lead to ineffective or even misleading advice during critical game moments.

To mitigate this, it's crucial to continually update AI models with data from varied playing styles, leagues, and game situations. Ensuring a diverse data set can help the AI provide more balanced and context-appropriate recommendations. Furthermore, integrating a mechanism for human oversight in AI decision-support tools can offer a safeguard, allowing coaches to weigh AI recommendations alongside their tactical knowledge and on-the-ground insights.

Ethical considerations play a significant role in addressing fairness and bias in AI. Sports organizations and technology developers must adopt ethical guidelines that prioritize fairness, inclusivity, and transparency. Establishing dedicated ethics committees or advisory boards can help monitor AI applications and ensure they align with these principles. These bodies can offer guidelines on data use, model development, and the continuous evaluation of AI systems for biases.

Education and awareness are also crucial. Coaches, players, and analysts must be trained not only in using these AI tools but also in understanding their limitations and potential biases. This understanding allows them to critically assess AI-generated insights and make informed decisions. Interactive workshops, seminars, and collaborative sessions can foster a more in-depth comprehension of AI's capabilities and constraints, promoting a more judicious use of these technologies.

Transparency in AI development and implementation also builds trust. Making the processes and decision-making frameworks of AI systems more transparent allows stakeholders to understand how conclusions are reached, the data used, and the potential biases that might arise. This transparency can be built through detailed documentation, open-source initiatives, or collaborative platforms where stakeholders can engage with the technology more directly.

Ultimately, balancing fairness and avoiding bias in AI approaches requires a multifaceted strategy. Rigorous data selection and continuous model evaluation are foundational steps. However, incorporating human judgment and maintaining ethical oversight are equally essential to ensure that AI serves the sport of field hockey in an equitable and beneficial manner. By addressing these ethical considerations proactively, the field hockey community can harness the full potential of AI while fostering a fair and inclusive environment for all players.

The Human Touch: Balancing AI and Human Judgment

Artificial Intelligence (AI) is undeniably transforming field hockey in myriad ways. From tailored training programs to real-time analytics during matches, the technology is making an indelible mark on how the game is played and coached. However, this raises an important question: How do we balance AI and human judgment to maintain the essence of the sport?

Field hockey, at its core, is a human endeavor colored by intuition, teamwork, and spontaneous creativity. While AI can analyze vast amounts of data and provide insights that were previously unimaginable, there are elements of the game that it can't fully encapsulate or predict. This is where the human touch becomes invaluable. Human coaches bring an innate understanding of their players' emotional and psychological states, something that even the most advanced AI systems struggle to replicate.

Moreover, coaches typically possess years of experience and a nuanced understanding of the game's dynamics that can't be easily translated into algorithms. For example, a player's sudden loss of confidence or a team's fluctuating morale in response to a series of unfortunate events on the field are aspects that AI might flag as anomalies but would fail to fully address. A seasoned coach, however, can read between the lines, providing the necessary guidance to lift the team's spirits, thereby impacting performance in ways AI predictions cannot.

The symbiosis of AI and human judgment offers a more holistic approach to the game. Consider real-time strategy adjustments; while AI can suggest optimal formations or tactical adjustments based on live data feeds, the final decision often rests with the coach. This is not merely a formality but a crucial step in ensuring that decisions align with the team's broader goals and unique dynamics. Coaches can weigh AI recommendations against their understanding of individual players, current morale, and the intuitive flow of the game.

In field hockey, adaptability is key. AI systems are typically trained on historical data, and while they can analyze patterns and make data-driven predictions, they aren't as adept at handling unprecedented events. Imagine a match scenario where weather conditions suddenly change or an unexpected injury forces a key player to leave the field. While AI might not immediately adapt to these outliers, a coach can quickly recalibrate the team's strategy to account for these unforeseen variables.

Even the aspect of leadership is something that remains quintessentially human. A coach provides not just tactical oversight but emotional support and moral leadership. When a team faces a nail-biting moment, it's the coach's pep talk, rather than any AI algorithm, that will galvanize the players and spur them to victory. This human element, rooted in leadership and emotional intelligence, is something AI currently cannot replicate.

Of course, it's not just about decision-making during matches. Human judgment plays a crucial role in interpreting AI-driven performance metrics. While AI can provide detailed analyses of a player's physical condition, skill execution, and other metrics, it's the coach and medical staff who contextualize these numbers into actionable insights. They understand that a player's performance on paper might not translate perfectly to real-world conditions due to myriad factors, including emotional well-being, fatigue, and even personal issues off the field.

This balance extends into other areas such as injury prevention and management. AI systems can predict potential injuries using complex algorithms and data from wearable technologies, offering a significant advantage in safeguarding athlete health. Yet, it's the medical professionals and coaches who interpret these predictions and make the final call on how to adjust training loads, rest periods, and recovery protocols. Their hands-on experience adds a layer of nuanced understanding that enhances AI's preventive capabilities

For young players, particularly those in the early stages of their careers, the mentorship and individualized feedback they receive from human coaches are irreplaceable. AI can support talent development through data-driven assessments and training recommendations, but the encouragement, inspiration, and life skills imparted by coaches are what often spark true growth and passion for the game.

There's also an ethical dimension involved in relying too heavily on AI. Decisions about player selection, game strategy, and even injury management inherently carry moral weight. Human judgment ensures that these decisions maintain a level of ethical integrity, balancing the utilitarian efficiency of AI with the ethical considerations every sport demands.

Furthermore, the world of field hockey thrives on unpredictability and excitement. Fans flock to the stands or tune into broadcasts to witness not just the skillful execution but also the human drama that unfolds in every match. The errors, triumphs, and split-second decisions made by human beings are what make the game engaging and relatable. An over-reliance on AI could render the game too predictable, stripping away some of the magic that makes field hockey a beloved sport worldwide.

Ultimately, the integration of AI in field hockey presents an incredible opportunity to enhance the game while highlighting the enduring importance of human judgment. When effectively balanced, AI can serve as an invaluable tool that augments human decision-making rather than replaces it. Coaches and players who leverage AI insights while also relying on their intuition, experience, and emotional intelligence will find themselves at a significant advantage.

The future of field hockey doesn't rest on choosing between AI and human judgment; it lies in harmonizing the two. As AI continues to evolve and integrate more seamlessly into the sport, the role of human judgment becomes even more critical. This balance ensures that field hockey remains not just a game of numbers but a rich tapestry of human experience, skill, and passion.

By fostering a partnership between AI and human judgment, we uphold the essence of the sport while also embracing the advancements that technology has to offer. This collaboration will not only elevate the level of play on the field but also deepen our appreciation for the unique human qualities that make field hockey the thrilling sport it is.

In conclusion, while AI offers a plethora of tools and insights capable of transforming field hockey, it's the human touch that rounds out this advanced approach. The best results come when both players and coaches blend data-driven insights with their instincts, experience, and emotional intelligence, ensuring that the beautiful game of field hockey remains ever captivating, ever human.

Chapter 10:
Case Studies and Real-World Applications

The transformative potential of artificial intelligence in field hockey becomes abundantly clear through various case studies and real-world applications. From the cutting-edge AI strategies employed by national teams to the grassroots level where local clubs harness machine learning algorithms for player development, these cases offer concrete proof of AI's impact. Notably, AI-driven predictive models have revolutionized strategic planning, helping coaches make smarter, data-backed decisions in real-time. Simultaneously, wearables and AI-enhanced video analytics are setting new standards in performance metrics and injury prevention. Learnings from other sports such as soccer and basketball also underline the universal applicability and adaptability of AI technologies. As these stories unfold, they don't just highlight successful implementations but also reveal invaluable lessons learned, pointing towards a future ripe with innovations. Moving forward, the challenge lies in scaling these applications while maintaining ethical considerations and balancing the indispensable human touch.

Successful Implementations of AI

Field hockey, like many other sports, has seen remarkable advancements through the integration of artificial intelligence (AI). Across the globe, teams, coaches, and analysts are leveraging AI to gain a competitive edge. From training enhancements to game strategy and performance analysis, successful implementations of AI in field hockey are heralding a new era of innovation and efficiency.

One notable example of successful AI implementation is in training programs. Teams use AI-powered systems to personalize training regimens for each player. By analyzing data from games, workouts, and even biometric sensors, AI algorithms can tailor exercises and drills to address specific strengths and weaknesses. This customization ensures that players receive the optimal level of training needed to enhance their performance.

In-game strategy has also benefited significantly from AI. For instance, real-time data analytics allow coaches to make informed decisions on the fly. AI-driven models can predict opponent moves and identify potential weaknesses in their strategy. This predictive capability enables teams to adapt swiftly during games, potentially turning the tide in their favor. The use of AI in this context isn't just about responding to the current state of play; it's about anticipating future scenarios and preparing accordingly.

Performance analysis has taken a leap forward with AI tools that offer detailed insights into every facet of a player's game. From tracking key performance indicators (KPIs) to video analysis, AI helps break down complex data into actionable insights. For instance, by analyzing video footage frame-by-frame, AI can identify subtle changes in a player's technique, helping coaches and players make precise adjustments. This level of scrutiny, previously only possible with extensive manual effort, is now automated and much more efficient.

The integration of AI in injury prevention and management represents another successful area of implementation. Wearable technology equipped with AI algorithms can monitor athletes in real-time, predicting potential injuries before they occur. By analyzing movement patterns, muscle exertion, and other variables, these systems alert players and coaches to signs of overuse or improper technique. Consequently, players can take preventive measures, thus reducing downtime and ensuring longer, healthier careers.

AI's impact extends beyond the players and onto the administrative and operational aspects of teams. Scouting and recruitment, for example, have been revolutionized by AI-driven scouting tools. These tools analyze vast amounts of data from various sources, including game statistics, social media activity, and even psychological assessments. The result is a more holistic view of potential recruits, helping teams identify future stars with greater accuracy.

Fan engagement is another arena where AI has successfully made its mark. AI in broadcasting and commentary provides real-time, insightful analysis that makes watching games more engaging. Enhanced interaction through technology, such as chatbots and social media, allows fans to connect with their favorite teams and players more intimately. Virtual and augmented reality experiences are also being developed, offering fans immersive ways to experience games from different perspectives.

While these implementations showcase the current capabilities of AI, they also set a precedent for future innovations. Sportscasters are increasingly relying on AI for better, faster, and more informative coverage. AI not only processes data at unprecedented speeds but also provides narratives and context, making the sport more accessible and enjoyable for viewers.

Field hockey teams worldwide are beginning to see the benefits of AI and early adopters are already reaping substantial rewards. For instance, the Belgium national team has incorporated AI into their training and strategy to great success, winning numerous accolades in recent years. Similarly, club teams in the Netherlands and Australia are using AI to refine their game strategies and player performance, setting new standards in their respective leagues.

However, the journey of AI in field hockey is not without its challenges. Implementing AI requires significant investment in technology and training. Teams must be willing to embrace a data-driven culture, which can sometimes be a shift from traditional methods. Moreover, the ethical considerations around data privacy and fairness must be addressed to ensure that the use of AI is both responsible and equitable.

Looking ahead, the potential for AI in field hockey is vast. Future innovations may include even more sophisticated predictive models, augmented reality tools for coaching, and AI-driven fan experiences that we can scarcely imagine today. Teams that continue to innovate and adapt to these technologies will undoubtedly find themselves at the forefront of the sport.

In conclusion, the successful implementation of AI in field hockey is a testament to the transformative power of technology. From enhancing training and game strategy to preventing injuries and engaging fans, AI offers numerous benefits that are reshaping the landscape of the sport. As AI continues to evolve, its role in field hockey will only grow, offering exciting prospects for players, coaches, analysts, and fans alike.

Lessons Learned from Other Sports

In examining the intersection of sports and artificial intelligence, it's crucial to recognize the invaluable lessons gleaned from various other sports which have successfully integrated AI into their practices. These instances offer a rich repository of insights, guiding field hockey towards making informed, strategic decisions.

Consider basketball, which has been a frontrunner in adopting AI technologies. The NBA's use of advanced analytics and machine learning algorithms to devise game strategies and player performance optimization stands as a hallmark of AI application. The sport leverages player tracking data to analyze movements, play efficiency, and even predict injury risks. Such practices have redefined the roles of coaches and analysts, making them increasingly data-centric. Field hockey can borrow from this by developing similar player tracking systems, refining training programs and game strategies based on real-time data analytics.

Soccer, or football as it is known globally, offers another instructive example. AI-driven tools like wearable GPS trackers and performance analysis software have transformed player and team performance assessments. These wearables collect vast amounts of data on player speed, acceleration, distance covered, and stamina. By mining this data, coaches can draw up tailored training regimens aimed at improving specific weaknesses and enhancing overall player fitness. For field hockey, employing similar wearable technologies can help in monitoring and optimizing player conditioning and tactics.

American football provides critical insights, especially on how predictive analytics can change the face of game strategy. The NFL's Next Gen Stats, powered by AI, offers unprecedented levels of detail about player movements, allowing coaches to make data-driven decisions. This system's ability to offer predictive analysis on play outcomes and player performances has added a new dimension to how the game is played and coached. Field hockey can implement these predictive models to enhance game strategies and anticipate opponent moves, thereby increasing the likelihood of successful outcomes.

When it comes to talent identification and recruitment, AI has revolutionized the approach in sports such as baseball. Major League Baseball (MLB) teams use AI-powered scouting tools to evaluate players based on a multitude of performance metrics, rather than relying solely on traditional scouting. Algorithms assess players' historical data to predict future performance, making the scouting process more objective and comprehensive. Field hockey could leverage similar technologies to scout talent across different levels, ensuring that the best players get noticed and nurtured efficiently.

Tennis showcases another innovative use of AI, particularly in injury prevention. For example, the use of AI to analyze video feeds and monitor players' physical stress levels helps in identifying potential injury risks before they become serious. This proactive approach allows for timely interventions, significantly reducing downtime due to injuries. Implementing such video analysis tools in field hockey can minimize injury risks and extend player careers.

One cannot overlook the significant contributions of AI in sports broadcasting and fan engagement, clearly demonstrated by cricket. Advanced AI algorithms are used to provide real-time statistics, player insights, and predictive analyses during live broadcasts. These innovations enrich the spectator experience, making games more engaging and informative. Field hockey can adopt similar technologies to enhance its spectators' experience, drawing in a larger and more engaged fan base.

Lessons from eSports are also worth mentioning, as they have set new standards in using AI for performance tracking and fan interaction. AI tools analyze in-game actions to provide feedback in real-time and simulate various scenarios for training purposes. They also engage fans through live chatbots and personalized content suggestions. Field hockey can look toward these practices to enhance both player training experiences and fan engagement.

Moreover, rugby's use of AI for game analysis and training drills is exemplary. AI systems analyze match footage to detect patterns and strategies used by opponents, helping teams to prepare more effectively for matches. These systems also aid in designing customized training drills that address team-specific needs and challenges. Field hockey can utilize similar AI methodologies to gain a comprehensive understanding of opponents' tactics and to develop precise training modules.

The integration of AI in sports such as swimming has shown how technology can transform performance timing and refinement. AI tools analyze minute details in swimmers' techniques and provide insights into how they can shave off precious milliseconds. For field hockey, this level of detailed analysis can boost player techniques and skills, ensuring that they optimize every moment on the field.

The sport of golf illustrates AI's impact on personalizing experiences and training. Golfers use AI-powered systems to analyze swing techniques, offering personalized feedback and suggesting improvements. Field hockey players could gain from such tailored insights to enhance their stick-handling, passing, and shooting techniques, driving individual and team performance.

Finally, the lessons drawn from these sports highlight the multi-faceted potential of AI. Whether it's through improved training, strategic game planning, injury management, or fan engagement, the benefits are manifold and undeniable. For field hockey, tapping into these insights means the opportunity to evolve and reach unprecedented levels of excellence. As AI continues to advance, the potential for further innovations is limitless, making it an exhilarating time for field hockey to embrace technology and revolutionize the game.

Future Prospects and Innovations

As we've seen in the previous sections, AI has already begun to revolutionize field hockey, and its potential for future innovations is immense. The rapid pace of technological advancements means that the way we train, play, and engage with the game could look drastically different in a few years. One of the most exciting prospects is the integration of AI with other emerging technologies like augmented reality (AR) and virtual reality (VR).

Imagine a training session where players can use AR goggles to receive real-time feedback on their technique, or a VR simulation that allows them to practice their skills in a hyper-realistic virtual environment. This isn't science fiction; it's a direction that the industry is actively exploring. Tech companies are already developing prototypes that merge these technologies with AI to offer unparalleled training experiences. These tools could help players improve faster and more efficiently than ever before.

Moreover, AI's ability to analyze massive datasets in real-time opens new frontiers in game strategy. Coaches could rely on AI not just for post-game analyses but also for making split-second decisions during matches. Integrating AI with live feed from the game could help in identifying patterns that are not immediately obvious, offering a competitive edge that is based on data-driven insights.

The future also holds promise in the realm of player health and safety. AI-driven predictive models for injury prevention are still in their infancy but show significant promise. By the time an athlete steps onto the field, they could be monitored in real-time for any signs of physical strain or potential injuries. This would allow for immediate interventions, possibly preventing injuries before they even occur. It also suggests a future where career longevity for players is significantly extended thanks to better management of their physical well-being.

Let's not overlook the potential for AI in talent discovery and management. AI has the capability to analyze the skills and attributes of young players across the globe, identifying future stars who might otherwise go unnoticed. Scout bots equipped with machine learning algorithms could revolutionize the recruitment strategies of clubs, making it easier to find and nurture the next generation of field hockey stars.

But innovation isn't just limited to the player's experience; it also involves enhancing the fan experience. AI and machine learning algorithms could make the game more interactive and engaging for fans. Personalized content delivery, AI-driven commentary, and interactive game features can bring the game closer to the audience. Imagine an AI commentator that gives insights tailored to your preferences, or a game app that allows you to interact with real-time statistics and player data during live matches.

In addition, integrating blockchain technology with AI can offer innovations in the way data is managed and utilized. Blockchain's secure and transparent nature dovetails perfectly with AI's data-driven needs, providing a secure environment for managing sensitive data. This is particularly crucial for maintaining the integrity of player statistics, game results, and historical archives.

Moreover, AI's capacity for continuous learning means that its models are always evolving. This means that the more it's used, the better and more accurate it becomes. It's conceivable that AI could soon offer predictive insights that are incredibly precise, helping teams get ahead of their competition by understanding trends and patterns that are beyond human capability to discern.

Looking further down the line, one might consider the ethical dimensions of these innovations. As much as technology promises to enhance the game, it also brings up questions about fairness, data privacy, and the role of human judgment. It's crucial that as we embrace AI, we also navigate its ethical implications responsibly. This includes ensuring data security, mitigating biases in AI algorithms, and maintaining a balance between AI inputs and human decisions.

Another fascinating domain is the application of AI in sports psychology. Understanding the mental state of athletes during games and practices can provide vital insights for performance enhancement. AI could be used to monitor stress levels, concentration, and overall mental health, offering valuable information that can be used to tailor training programs to meet the psychological needs of each athlete. This holistic approach could help in building not just better players but more resilient individuals.

Similarly, AI could play a role in improving field facilities themselves. Smart fields equipped with sensors and AI systems could offer optimal playing conditions by adjusting factors like turf maintenance, lighting, and even weather preparedness. This would ensure the best possible environment for both training and competitive play, minimizing external variables and supporting peak performance at all times.

The commercialization of AI technologies in field hockey opens another avenue for innovation. As AI tools become more sophisticated and accessible, there's potential for new business models and revenue streams. Companies specializing in AI solutions for sports could offer subscription-based services for teams at all levels, from amateur to professional. This could democratize access to advanced analytics and training tools, making high-tech solutions available to a broader audience.

As we look to the future, the integration of AI in field hockey is likely to be a collaborative effort. Universities, tech companies, sports organizations, and even startups are all playing a part in pushing the boundaries. This collaboration fosters a fertile ground for groundbreaking innovations that we might not have even imagined yet. Partnerships between these entities can accelerate the development and deployment of AI solutions, making the future arrive sooner than we think.

Fans, too, have a role to play in shaping these innovations. As the end-users of enhanced viewing experiences, their feedback will be invaluable in refining AI applications for fan engagement. The dynamic between fans and technology is symbiotic, each driving the other towards greater heights of engagement and enjoyment. AI can offer tailored experiences, but the real magic happens when fans interact with these technologies, providing data and insights that further improve the system.

In conclusion, the future of AI in field hockey is not just about what the technology can do, but about how it can reshape our experience of the sport. It's about making training more effective, games more strategic, and the overall experience—from players to fans—more immersive and enjoyable. We're on the cusp of a new era in field hockey, where technology and talent combine to create a game that's not just played, but lived. The innovations on the horizon promise to take the sport to new heights, creating a more dynamic, engaging, and intelligent future for field hockey.

Chapter 11:
Integrating AI into Field Hockey Programs

As the field hockey world increasingly embraces artificial intelligence, integrating AI into existing programs requires a strategic approach that maximizes benefits while minimizing disruption. It's essential to foster a culture open to technological advancement, beginning with targeted training sessions for coaches and players to fully understand AI-driven tools. These educational initiatives ensure that everyone involved can leverage AI for personalized training and advanced game strategies. Additionally, carefully curated implementation steps, starting from pilot programs and scaling up, provide a roadmap for seamless integration. Finally, continuous evaluation and feedback loops help refine these AI applications, ensuring consistent improvement and tangible success. By embedding AI thoughtfully into the core of field hockey programs, the sport can reach unprecedented levels of performance, strategy, and engagement.

Steps for Implementation

Integrating AI into a field hockey program requires careful planning, dedication, and a deep understanding of both the sport and the technology. Here's a breakdown of actionable steps to ensure a seamless integration process.

Step one is identifying the areas where AI can make the most substantial impact. This involves comprehensive consultations with coaches, players, and analysts to pinpoint current challenges and opportunities for improvement. Common areas include performance tracking, strategy development, and injury prevention. By understanding the specific needs and goals of the team, it's easier to match the right AI tools and technologies to the problem.

The next step is conducting thorough research on available AI solutions. The field of AI is vast and evolving rapidly, so it's crucial to stay updated on the latest advancements. This might involve attending industry conferences, participating in webinars, and reading up-to-date literature. The goal is to gather a broad spectrum of potential tools and applications. This research phase also includes evaluating the scalability, compatibility, and ease of use of the AI systems under consideration.

Once the research phase is completed, the focus shifts to pilot testing. Start small—perhaps with a specific aspect of training or a single team unit. This pilot phase allows for experimentation and adjustment without the risk of widespread disruption. Gather feedback from all stakeholders involved in the pilot, including coaches, players, and support staff. Use this feedback to tweak and enhance the AI system before a full-scale implementation.

Training is the cornerstone of successful AI integration. Coaches, players, and analysts must be well-versed in using AI tools. This training should be hands-on and continuous, focusing not just on the operational aspects but also on understanding the data outputs and making informed decisions based on AI insights. Training sessions can include workshops by AI experts, online courses, or in-house training facilitated by the technology providers.

Creating an interdisciplinary team is essential. Integrate AI specialists with the existing coaching and support team to ensure that technical expertise aligns with practical field knowledge. This team will act as the bridge between the technology and its application in training, strategy, and game performance.

Data management is another critical step. AI thrives on data, and the quality, accuracy, and accessibility of this data are paramount. Establish protocols for data collection, storage, and analysis. Use standardized formats and ensure data privacy and security. Developing a centralized data repository can help streamline access and analysis, making it easier to draw actionable insights.

Regularly evaluate and tweak the AI systems in place. AI technologies are not static; they improve and adapt over time. Continuous evaluation helps ensure that the AI tools remain effective and relevant. This might involve regular performance audits, feedback sessions, and updates from technology providers.

Communication plays a crucial role throughout the implementation process. Keep all stakeholders informed about the progress, challenges, and successes of the AI integration. Transparent communication fosters trust and ensures everyone moves towards a common goal. Regular meetings, reports, and updates can keep everyone aligned and motivated.

The issue of cost and resource allocation cannot be ignored. Integrating AI into a field hockey program requires financial investment and resources. Budget considerations should cover not just the acquisition of technology but also training, maintenance, and potential upgrades. Allocating resources wisely can help in achieving a balanced approach where the benefits of AI are maximized without overwhelming financial constraints.

Lastly, celebrate milestones and successes. Integrating AI is a significant step forward, and recognizing the efforts and achievements can boost morale and encourage further innovation. Whether it's a small win in improved player performance or a major game strategy breakthrough, acknowledging these milestones helps reinforce the positive impact of AI.

In summary, the integration of AI into field hockey programs is a multifaceted process that involves careful planning, research, pilot testing, training, data management, continuous evaluation, clear communication, resource allocation, and celebration of successes. Each step is integral to ensuring that AI becomes a valuable asset rather than a disruptive force, enhancing the game and reaping the benefits of this advanced technology.

Training for Coaches and Players

Integrating AI into field hockey programs involves more than just installing software and collecting data; it's about transforming the way coaches and players approach the game. The main goal is to ensure that both coaches and players understand how to leverage artificial intelligence to enhance their skills, decision-making processes, and overall performance.

A critical first step in this transformation is educating coaches about the potential of AI. Many coaches might be accustomed to traditional methods and hesitant to adopt new technologies. However, AI offers the ability to analyze vast amounts of data quickly and accurately, providing insights that would be nearly impossible to gather manually. Training sessions for coaches should focus on how to interpret AI-generated reports, utilize real-time analytics, and incorporate these insights into training routines and game strategies.

For players, the introduction of AI means a shift from generic training programs to highly personalized ones. AI can assess individual player strengths and weaknesses in unparalleled detail. For instance, wearable technology and smart equipment can track a player's movements, stamina, speed, and even stress levels. By analyzing this data, AI can generate customized training regimens that target specific areas of improvement. This type of targeted training leads to faster skill development and helps players reach their peak performance levels more efficiently.

One practical application of AI in training is the use of computer vision and machine learning to break down and analyze game footage. Coaches can use AI to highlight patterns and trends in both their team's performance and that of their opponents. These patterns can reveal weaknesses in defense, predict opponent strategies, or identify opportunities for offensive plays. Armed with this information, coaches can devise more effective game plans and tailor their training sessions to address specific tactical needs.

Moreover, AI-powered tools can simulate various game scenarios, allowing players to practice responses to different in-game situations. This kind of simulation is invaluable for developing quick decision-making skills on the field. For example, a player could practice taking penalty corners in a virtual environment that mimics the pressure and unpredictability of a real match, preparing them better for actual game conditions.

Besides technical and tactical training, AI can play a significant role in mental and psychological preparation. Programs that monitor physiological data such as heart rate variability and cortisol levels can indicate a player's stress and fatigue levels. Coaches can use this information to tailor mental conditioning programs that help players manage stress and improve focus and resilience. Incorporating these psychological insights into training not only enhances player performance but also contributes to overall well-being.

Another exciting dimension is the collaborative aspect of AI tools. Platforms that facilitate communication and data sharing between players and coaches can enhance the training experience. Imagine a system where each player has their performance dashboard that they can access anytime to review their stats, watch customized video clips of their plays, and get recommendations for improvement. This transparency and accessibility empower players to take responsibility for their development while keeping coaches informed and engaged.

Injury prevention and management is another critical area where AI can make a substantial difference. AI models that predict injury risks based on player data can inform training schedules, ensuring that players are neither overtrained nor undertrained. For instance, if the data suggests that a player is at risk of a hamstring injury, the AI system can recommend modifications to their training routine to mitigate this risk. Early interventions can reduce the likelihood of injuries, keeping players fit and available for selection.

Implementing AI in training requires a cultural shift within the team. Everyone involved, from the coaching staff to the players, needs to be open to adopting new technologies and methods. It's not just about learning how to use AI tools, but also about integrating them seamlessly into everyday practices and routines. Continuous education and support from AI specialists can ease this transition, ensuring that the team fully utilizes the technology.

While AI offers numerous benefits, it's essential to maintain a balance between technology and the human touch. The intuition, experience, and emotional intelligence of coaches and players are irreplaceable. AI should be seen as an ally that complements and enhances human capabilities rather than a substitute. Effective integration of AI means blending data-driven insights with real-world expertise to create a comprehensive and dynamic training environment.

Furthermore, the dynamic nature of AI means that training for coaches and players must be ongoing. As AI technologies evolve, so too must the strategies and tactics employed by the team. Regular workshops and refresher courses can help keep everyone updated on the latest advancements and best practices. Staying ahead of the curve will ensure that the team remains competitive and maximizes the benefits that AI offers.

Lastly, it's crucial to evaluate the effectiveness of AI integration regularly. Metrics and KPIs can help in assessing whether the AI-driven training programs are delivering the desired results. Are players showing measurable improvements in their skills and fitness? Are the strategies developed through AI analytics leading to better in-game performance? Regular evaluations will help in fine-tuning the AI applications and ensuring that they remain relevant and effective.

Integrating AI into field hockey training is not just a short-term project but a long-term commitment to innovation and excellence. As coaches and players embark on this journey, the potential to redefine what's possible in the sport becomes an exciting reality. By embracing AI, the field hockey community can unlock new levels of performance, strategy, and enjoyment of the game.

Evaluating Success and Improvements

Integrating AI into field hockey programs demands not just an initial investment but continuous assessment to maximize its efficacy. The ability to evaluate the success and areas for improvement is pivotal to ensure that AI tools deliver on their promises and contribute positively to both player performance and game strategies. In assessing the success of AI integration, a multifaceted approach is essential, encompassing key performance indicators (KPIs), feedback mechanisms, and iterative refinements.

First and foremost, clearly defining KPIs is a crucial step. For trainers and coaches, these indicators might include metrics such as improved player statistics, reduced injury rates, and enhanced game strategies. Specific KPIs could be as granular as the number of successful passes in a game or the reduction in recovery time after injuries. Establishing a baseline before the integration of AI systems to measure these KPIs against is an invaluable part of the evaluation process.

Feedback from players and coaches is another significant aspect of evaluating the success of AI integration. Anecdotal evidence and qualitative data gathered through regular interviews, surveys, and feedback sessions offer insights that raw data might miss. It's important to foster an environment where players and coaches feel comfortable sharing their experiences and suggestions for improvements. This feedback mechanism allows for a holistic understanding of the tools' impact beyond just numbers.

Consider the example of using AI-driven wearable technology to monitor player health and performance. Feedback might reveal that while the technology accurately tracks data, it could be cumbersome to wear or distracting during practice sessions. Such feedback is critical for iterative improvements. Developers can then tweak the design or functionality to better suit user needs, thus enhancing overall effectiveness.

Regular performance reviews and audits of AI systems are also paramount. These audits should not be limited to technical checks but must also consider practical applications and results on the field. Is the AI tool helping players react faster during matches? Is it providing analytics that adapt smoothly to real-time changes? Reviewing these aspects can unveil gaps and areas of improvement to focus on going forward.

Importantly, collaboration with AI experts and developers furthers the refinement process. Engaging with those who develop and maintain the AI systems ensures that issues are promptly addressed, and upgrades are continually implemented. This close collaboration helps keep the AI tools at the cutting edge, adapting to new insights and technological advancements seamlessly.

Looking at how AI has reshaped training regimes can offer insightful case studies. For instance, AI tools that analyze player movements to refine techniques must demonstrate measurable improvements in skills over time. Video analysis powered by AI can be compared to traditional coaching methods to determine its incremental benefits. These reflections provide a comparative framework to evaluate the merit and necessity of AI in training versus conventional techniques.

Furthermore, field hockey programs should incorporate a phased approach to implementing AI. Starting with pilot projects allows for initial testing and adjustment before a full-scale rollout. This methodical approach helps identify potential pitfalls and areas needing adaptation early on. Pilot projects serve as miniature case studies from which larger programs can extrapolate lessons and best practices.

One can't overlook the need for continuous education and training for both players and coaches. AI tools are only as effective as the users operating them. Ensuring that all team members are proficient in using these technologies goes a long way in maximizing their benefits. Regular workshops, hands-on training sessions, and access to online resources provide an ongoing learning environment, helping teams stay updated with the latest advancements and techniques.

Real-time analytics offer another layer of real-world application that requires constant monitoring and improvement. While predictive models and AI-driven strategies can optimize decision-making, their ultimate success hinges on real-match scenarios. Are these tools providing insights that translate to better on-field performance? How quickly and accurately do these systems adapt to the unpredictability of live matches? Monitoring these aspects over multiple games is necessary to assess their true value.

Moreover, integrating AI isn't a one-time effort; it demands continuous improvements and updates. AI systems learn and evolve, but they need curated data and regular fine-tuning to maintain effectiveness. Incorporate regular update cycles wherein the AI tools are recalibrated based on recent data and developments. This ongoing improvement cycle is vital to ensure that AI tools remain relevant and continue to offer a competitive edge.

Another aspect worth considering is the resource investment required. Evaluating the financial cost versus the gained benefits is part of a thorough assessment. While AI has the potential to revolutionize training and game strategies, its cost-effectiveness must be justified by tangible improvements in performance and outcomes. This includes assessing long-term savings, such as reduced injury-related expenses due to better monitoring and rehabilitation guided by AI tools.

Finally, documenting successes and improvements offers a transparent way to track progress over time. Comprehensive reports detailing the performance metrics pre- and post-AI integration, feedback summaries, case studies, and improvement logs create a robust repository of evidence. This documentation not only serves internal review purposes but also provides a framework for communicating successes to stakeholders, potential sponsors, and the wider field hockey community.

In summary, evaluating the success and areas for improvement of AI in field hockey programs is a detailed and holistic process. By establishing clear KPIs, fostering feedback mechanisms, conducting regular performance reviews, engaging in continuous education, and maintaining a cycle of iterative improvements, teams can ensure that AI tools live up to their potential. This meticulous approach not only aids in leveraging AI for current benefits but also sets a foundation for future advancements and sustained success.

Chapter 12:
The Future of AI in Field Hockey

The integration of AI into field hockey is just beginning to scratch the surface, revealing a horizon teeming with possibilities. Emerging trends like AI-driven tactical analysis and advanced machine learning algorithms promise to refine every aspect of the game, from player conditioning to real-time decision-making on the field. As these technologies evolve, they will arm coaches and players with unprecedented insights, driving performance to new heights. The long-term implications extend beyond merely enhancing individual elements; they herald a holistic transformation in how the sport is played, watched, and understood. Preparing for this AI-driven future means embracing ongoing education and adaptability from all stakeholders—ensuring that field hockey not only keeps pace with technological advancements but leads the charge in innovation.

Emerging Trends and Technologies

As artificial intelligence continues to weave itself into the fabric of our daily lives, it's no surprise it's making waves in field hockey. The sport, traditionally driven by physical prowess and tactical acumen, is now experiencing a digital revolution. Emerging trends and technologies are not just improving the game but also redefining how it is played, coached, and experienced. These innovations offer a glimpse into a future where AI's role becomes increasingly integral, profoundly impacting every facet of field hockey from training to gameplay.

One of the most exciting emerging trends in AI is the use of machine learning algorithms to analyze player performance in real-time. These algorithms can process vast amounts of data quickly, providing insights that were previously unimaginable. AI can track a player's movements, identify patterns, and offer real-time feedback to coaches and players. This not only enhances immediate decision-making during games but also aids in developing strategic training regimens tailored to individual players' needs and abilities.

Wearable technology is another burgeoning field that promises to revolutionize field hockey. Devices equipped with sensors can monitor a player's heart rate, speed, distance covered, and even biomechanical data points such as joint angles and muscle activation. This information is then fed into AI systems that analyze the data, offering insights that can help prevent injuries and optimize performance. Such technology empowers players and coaches to focus on areas that need improvement, ensuring a more scientific approach to training and recovery.

In the realm of tactical planning, AI is starting to play a crucial role. Coaches can employ predictive modeling to simulate different game scenarios and determine the most effective strategies. These models leverage historical data and machine learning to predict the outcome of various tactical decisions. The insights garnered from these simulations enable coaches to fine-tune their strategies, making informed decisions that can significantly affect the outcome of matches.

The integration of AI into video analysis tools is another trend that's gaining traction. Advanced video analysis platforms can dissect match footage to identify key performance metrics and tactical insights automatically. By recognizing patterns that human eyes might miss, these tools offer a detailed breakdown of both individual and team performances. Coaches can then use these insights to address weaknesses, capitalize on strengths, and develop highly targeted training sessions.

Furthermore, the development of AI-driven coaching tools is setting new standards in personalized training. These tools offer a bespoke training experience by analyzing an individual player's strengths and weaknesses. Through tailored drills and exercises, players can work on specific areas needing improvement. This individualized approach ensures that every player, regardless of their skill level, can benefit maximally from training sessions.

AI is also making strides in fan engagement and experience. Augmented reality (AR) and virtual reality (VR) technologies are transforming how fans interact with the game. By offering immersive experiences, fans can feel closer to the action, even if they are miles away from the pitch. These technologies can overlay real-time stats and analytics during live broadcasts, providing enthusiasts with deeper insights into game dynamics. The ability to visualize complex data in an engaging manner not only enhances the viewing experience but also attracts a broader audience base.

Another promising area is the application of AI in injury prevention and management. Predictive models can analyze a player's biomechanics and physical condition to predict the likelihood of injuries. By identifying potential risks early, preventative measures can be taken to mitigate these risks. This proactive approach helps maintain player health and longevity, ensuring that athletes can perform at their best over extended periods.

In recruitment and talent identification, AI is starting to change the game. Scouting tools are becoming more sophisticated with AI's ability to analyze and compare player metrics across various leagues and competitions. Talent assessment has never been more data-driven, allowing scouts to identify promising players that might have been overlooked through traditional scouting methods. AI can analyze vast amounts of data, from physical attributes to psychological profiles, ensuring a thorough evaluation of potential recruits.

On the ethical front, as AI becomes more embedded in field hockey, considerations around data privacy and fairness are paramount. It's essential to ensure that AI systems are transparent and that the data used is secure and responsibly managed. Equally important is addressing the inherent biases in AI algorithms to ensure fair and equitable outcomes. Balancing the efficiency of AI with the human touch is crucial to maintaining the integrity of the sport.

AI's role in field hockey is not just limited to the elite levels of the game. Grassroots programs can also benefit immensely from these technologies. From talent identification to tailored coaching programs, AI can provide tools and resources that were previously out of reach for smaller clubs and communities. This democratization of technology ensures that the future stars of the game can emerge from any background, paving the way for a more inclusive sport.

As we look to the future, the integration of AI into field hockey is poised to grow even deeper. The advancements we see today are just the tip of the iceberg. Future innovations could include AI referees to assist or even replace human officials, ensuring fairer outcomes through impartial decision-making. Autonomous drones could provide new angles for game footage, offering unparalleled insights and enhancing the fan experience further.

The long-term implications of AI in field hockey are profound. With continuous advancements, AI systems will become more intuitive and seamlessly integrated into every aspect of the game. The sport will evolve, becoming faster, smarter, and more strategic. Players will reach new heights of performance, driven by data and insights previously beyond their grasp. Coaches will transform their strategies, guided by the precise understanding that AI offers. And fans will

Hockey and AI

engage with the game in ways that are more immersive and connected than ever before.

In preparation for this future, it's essential for all stakeholders in field hockey to embrace these emerging technologies. Training programs for coaches and players must include AI literacy to ensure that everyone can make the most of these powerful tools. Investment in technology should be a priority for clubs looking to stay competitive, and collaborations between tech companies and sports organizations will drive further innovation.

Ultimately, the future of AI in field hockey is one of promise and potential. By harnessing the power of AI, the sport can evolve in ways that were once unimaginable, offering richer experiences for players, coaches, and fans alike. As we stand on the brink of this technological revolution, it's an exciting time to be involved in field hockey, with AI leading the charge towards a new era of innovation and excellence.

Long-Term Implications

The future of AI in field hockey isn't just a fleeting trend; it's a seismic shift that holds profound implications for the sport's long-term trajectory. As artificial intelligence becomes more integral to how the game is played, managed, and consumed, we need to consider what this means for everyone involved. From players and coaches to fans and governing bodies, the ripple effects are vast and varied.

Field hockey is traditionally a sport rooted in physical prowess and strategic acumen. However, as AI-driven tools integrate into every facet, from training sessions to live matches, the essence of competition may undergo a transformation. Take, for example, the way data analytics and predictive modeling can offer an edge previously unimaginable. Coaches armed with AI can not only fine-tune strategies in real-time but also predict opponent actions with surprising accuracy. This level of precision can fundamentally alter match outcomes and redefine the very concept of "home field advantage."

For players, the impact of AI extends far beyond the pitch. Long-term career development will likely rely on more personalized, data-driven insights into their performance. Imagine an AI that tracks a player's growth over years, identifying muscle imbalances or technique flaws before they result in injury. Such technologies might not only prolong careers but also elevate the overall quality of play.

However, the implications don't stop at the professional level. Grassroots and youth development programs will also reap benefits. AI tools can democratize high-level training, making sophisticated assessments and tailored programs accessible to young talent worldwide. This level of access can level the playing field, potentially unearthing talent from unexpected quarters and diversifying the sport.

While the potential gains are substantial, we must also consider the ethical complexities AI introduces. Automated decision-making systems bring questions around fairness, transparency, and bias. Algorithms trained on biased data could perpetuate existing inequities, disadvantaging certain players or teams. Steps must be taken to ensure that AI augments human decision-making and doesn't supplant it, preserving the human touch that makes sports vibrant.

Injury prevention and management is another area ripe for transformation through AI. Predictive injury models, wearable technology, and AI-driven rehabilitation protocols could revolutionize how players stay fit and recover. In the long-term, this might significantly reduce the physical toll on athletes, potentially extending their professional lives and enhancing the game's overall level of play.

Fan engagement will also see dramatic changes. AI technologies can create more interactive and personalized fan experiences, from enhanced broadcasting to virtual and augmented reality. This brings sports fandom into a more immersive era, though it might also change how fans interact with and feel about the sport. Traditional experiences such as attending games in person may take on new dimensions when augmented with these technologies.

Regulating bodies and leagues will need to adapt as well. The introduction of AI calls for new rules and policies to govern its use, balancing technological advancement with the sport's integrity. Regulations around data privacy, fairness, and even performance-enhancing technologies will become hot topics, requiring thoughtful dialogue and judicious rules-making.

In the educational realm, field hockey coaching needs a paradigm shift to prepare for this AI-infused future. Coaches will require training not only in traditional methods but also in leveraging advanced technological tools. This dual expertise will be crucial for those aiming to stay competitive in the evolving landscape.

From a macro perspective, AI in field hockey could spur innovations that bleed into other sports and industries, creating a virtuous cycle of technological advancement. Collaborative efforts between tech companies and sports organizations can bring about innovations that no single entity could achieve alone, fostering an ecosystem where mutual growth thrives.

Yet, with all these prospects, there's an undercurrent of caution. The AI-driven future must be approached with a responsible mindset, considering the long-term societal impacts. Over-reliance on technology could lead to a diluted form of the sport, where human intuition and spontaneity are undermined. Thus, while embracing AI, stakeholders must strive for a balanced integration that preserves the heart and soul of field hockey.

In sum, the long-term implications of AI in field hockey are both exciting and complex. They offer opportunities for unprecedented advancements in training, performance, and fan engagement while prompting essential discussions around ethics, regulation, and the human element. As we stand on the brink of this transformation, it's not just about leveraging AI for short-term gains; it's about shaping a future where the sport flourishes in harmony with technological progress.

Preparing for the Future

The landscape of field hockey is rapidly evolving, with artificial intelligence playing an increasingly pivotal role. Players, coaches, and analysts must continually adapt to stay ahead. To prepare for this future, it's essential to understand the emerging trends and technologies, integrate them effectively, and anticipate their long-term implications.

Let's start by acknowledging that preparation involves both mental and technological readiness. Players and coaches need to shift their mindset towards a more data-driven and analytical approach. Embracing AI means rethinking training regimes, game strategies, and performance evaluations.

For field hockey players, keeping up with AI-enhanced training tools is only part of the journey. It also means being open to continuous learning. AI provides personalized feedback tailored to each player's strengths and weaknesses, making it crucial to absorb and act on this feedback. Taking advantage of AI's capacity to offer minute adjustments, players can refine their skills with a precision that was previously unattainable.

Coaches, on the other hand, must become savvy in utilizing AI-driven data analytics and insights. They need to interpret this data effectively and integrate it into their training and game strategies. Skills once rooted in instinct and experience now require a balance of intuition and data science. By leveraging AI, coaches can anticipate opponent moves, optimize player formations, and fine-tune their strategies in real-time.

Performance analysts have a crucial role in this evolving scenario. Being at the confluence of data science and sports strategy, they must stay updated with the latest AI technologies. Proficiency in machine learning algorithms and deep learning models will be essential. Crucially, analysts must learn to differentiate between meaningful insights and data noise, ensuring that AI contributes positively to decision-making processes.

But preparing for the future isn't just about technology; it's also about building a robust infrastructure that can support AI applications. Clubs and organizations need to invest in high-quality data collection tools such as advanced sensors, state-of-the-art cameras, and wearable technology. This ensures that the data fed into AI systems are accurate and representative.

Beyond the hardware, there's a need for robust software solutions that can process and analyze this data. Teams must partner with AI developers to create custom solutions tailored to their specific needs. Off-the-shelf solutions might not offer the precision and personalization necessary for optimal performance.

An often-overlooked aspect is educating the stakeholders. Coaches and players should be trained not just in using these AI tools but also in understanding how they work. This demystification of AI can foster a more collaborative environment where technology complements human intuition rather than competing with it.

Ethics and data privacy should also be cornerstones of future preparations. Ensuring that player data is used responsibly and securely is critical. AI's integration in field hockey must be guided by principles that prioritize fairness, transparency, and the well-being of all participants.

Another influence to consider is the culture within teams and organizations. Cultivating a culture that embraces technology while valuing traditional skills and knowledge is essential. It's not about replacing coaches or analysts but augmenting their capabilities. The future of AI in field hockey lies in harmonious integration.

Looking further, field hockey's governing bodies and leagues must play an active role in setting standards and guidelines for AI usage. Establishing best practices and creating a framework for integrating AI across different levels of the sport will ensure consistency and fairness. They need to be proactive in addressing potential issues like bias in AI systems and ensuring equitable access to these technologies.

Moreover, collaboration between different stakeholders will drive progress. Clubs, universities, and tech companies must work together to conduct research, develop new AI applications, and share their findings. This collaborative approach can accelerate the development of innovative solutions and ensure they are tested and refined in real-world scenarios.

In the longer term, the implications of AI in field hockey could be profound. As algorithms become more sophisticated, they may uncover new facets of the game that were previously hidden. For instance, AI could identify previously unnoticed patterns, leading to the development of new training techniques or strategic innovations. These advances will continuously reshape our understanding of the game.

Preparation also involves being adaptable to unexpected changes. As AI evolves, new challenges and opportunities will emerge. Those who can pivot and adapt quickly will have a competitive edge. Therefore, fostering a mindset of flexibility and continuous learning is pertinent.

Field hockey's future with AI isn't a distant vision; it's quickly becoming the present. Preparing for this future means embracing a blend of technology and tradition. It means staying curious, continually learning, and leveraging AI to improve every aspect of the game. For those ready to embrace this change, the future doesn't just look promising—it looks revolutionary.

With a keen focus on innovation, a grounded understanding of the game, and a commitment to ethical practices, the field hockey community can navigate this transformative era. The fusion of AI and field hockey is set to chart new territories, redefine excellence, and inspire the next generation of players and coaches. The journey is underway, and the field is wide open for what comes next.

Conclusion

We've journeyed through the intricate tapestry woven by artificial intelligence and field hockey, witnessing firsthand how AI is not merely an accessory but a cornerstone in the modern evolution of the sport. It's evident that as AI continues to integrate within the game, the possibilities are boundless. From reshaping training programs to redefining game strategies, and from enhancing player safety to enriching fan experiences, the landscape of field hockey is undergoing a remarkable transformation.

AI's influence in training is arguably where the immediacy of change is most tangible. Personalized programs and AI-driven coaching tools have taken individualized player development to a level previously unimaginable. These advancements mean that every player's unique strengths and weaknesses can be identified and honed, fostering a new generation of athletes capable of pushing the boundaries of what is possible on the field. Comprehensive skill development and refinement facilitated by AI are setting the stage for a sharper, more competitive era in field hockey.

The strategic dimension of field hockey is also experiencing a paradigm shift due to real-time data analytics and predictive modeling. The ability to analyze vast amounts of data instantaneously offers teams an unprecedented edge, strategically positioning themselves to capitalize on every opportunity. Coaches now have at their disposal a suite of AI-assisted decision-making tools that allow for more informed and agile responses during high-stakes matches. This synthesis of human intuition and AI precision is crafting a more dynamic and unpredictable game, much to the delight of fans and players alike.

Performance analysis, enriched by AI tools and video analysis, provides a granular view of every aspect of an athlete's gameplay. Key performance indicators that were once laborious to track manually can now be monitored seamlessly, offering insights that are both deep and actionable. The calibration of performance metrics introduces a level of objectivity and clarity, enabling players and coaches to measure improvements with pinpoint accuracy.

Injury prevention and management are other critical areas where AI is making substantial inroads. Wearable technology and predictive injury models are helping to safeguard the health and longevity of athletes. These innovations allow for proactive interventions that can prevent injuries before they occur and offer tailored rehabilitation processes that expedite recovery. The integration of AI in this sphere is not just advancing athletic performance but also ensuring that athletes can enjoy longer, healthier careers.

When it comes to recruiting and talent identification, AI's role can't be overstated. The precision of scouting tools and techniques, along with AI's ability to assess talent based on quantifiable metrics, is revolutionizing how new players are discovered and groomed. This technology is making the scouting process both more equitable and efficient, uncovering hidden talents and providing them with opportunities that might have otherwise been missed. The development of future stars is being accelerated, ensuring that the sport remains vibrant and competitive.

The fan engagement experience has transformed dramatically with AI enabling more interactive and immersive experiences. AI-driven technologies in broadcasting and commentary are bringing fans closer to the action, providing a richer, more intuitive viewing experience.

Augmented and virtual reality are adding layers of depth to fan interaction, creating a participatory ambiance that was unimaginable just a few years ago. This enhanced connectivity is cultivating a more engaged and passionate fanbase, driving the sport's popularity even further.

Ethical considerations are at the forefront as we integrate AI into field hockey. Balancing data privacy, fairness, and transparency with the undeniable benefits AI brings is a tightrope walk that demands conscientious oversight. Ensuring that AI technologies are implemented responsibly is crucial to maintaining the integrity of the sport. The human element, the soul of field hockey, must remain intact even as we embrace technological advancements. A symbiotic relationship between human judgment and AI will always be vital.

Real-world applications and case studies underscore the profound impacts AI has already had across various sports, offering valuable lessons and insights that field hockey can leverage. These examples serve as both inspiration and a roadmap, highlighting successful implementations and the potential pitfalls to avoid. The continuous evolution in AI technologies promises even greater innovations, emphasizing the importance of staying abreast with emerging trends to harness their full potential.

Integrating AI into field hockey programs is a multi-faceted process requiring careful planning and execution. From training coaches and players to evaluating the success and improvements, each step is crucial for seamless adaptation. The ultimate objective is to create an environment where AI and human expertise coalesce to elevate the game to new heights. Monitoring and refining these implementations will ensure that AI's contributions are both effective and sustainable.

Looking ahead, the future of AI in field hockey holds tremendous promise. Emerging trends and long-term implications suggest a sport continually pushing the envelope of innovation. Preparing for this future involves not just adapting to change but anticipating it. By staying forward-thinking and flexible, the field hockey community can ensure it thrives in an era where AI is an integral part of the game's fabric.

In conclusion, the transformative impact of AI on field hockey is undeniable and ubiquitous. It offers tools and capabilities that are reshaping how the game is played, coached, analyzed, and experienced. However, with great power comes great responsibility. The onus is on us to wield this power wisely, ensuring that while we embrace technological advancements, we preserve the core spirit and integrity of the sport. By doing so, we not only honor the rich legacy of field hockey but also pave the way for a future where the game can flourish in new and exciting ways.

As we stand on the cusp of this exciting frontier, it's clear that the fusion of AI and field hockey promises a thrilling journey filled with endless possibilities. With AI as a catalyst, the evolution of this beloved sport will continue, forging a path that is both innovative and inspiring.

Thank you for embarking on this journey through the intersections of artificial intelligence and field hockey. The future awaits us, full of promise and potential.

Appendix A:
Appendix

This appendix serves as a vital resource for supplementing the content discussed throughout the book. It provides additional context, references, and detailed explanations that further elucidate the transformative impact of artificial intelligence on field hockey. Readers will find expanded discussions, charts, and data sources that are crucial for a deeper understanding of the material. Whether you're a player, coach, analyst, or simply a sports enthusiast, the appendices offer pragmatic insights, additional reading lists, and practical tools to apply AI technologies effectively. This section ensures that all readers, regardless of their entry-level knowledge, can navigate through the intersections of sports and technology with greater clarity and purpose.

Glossary of Terms

This glossary serves as a reference to understand the key terms used throughout the book on how artificial intelligence (AI) is transforming the world of field hockey. Understanding these terms will help you grasp complex concepts and appreciate the technological advancements discussed in the chapters.

Algorithm: A set of rules or procedures for solving a problem in a finite number of steps, often used by computers to process data.

Artificial Intelligence (AI): The simulation of human intelligence processes by machines, especially computer systems, including learning, reasoning, and self-correction.

Augmented Reality (AR): An enhanced version of reality created by the use of technology to overlay digital information on an image of something being viewed through a device.

Big Data: Extremely large data sets that can be analyzed computationally to reveal patterns, trends, and associations, especially relating to human behavior and interactions.

Data Analytics: The process of examining data sets to draw conclusions about the information they contain, often with the aid of specialized systems and software.

Deep Learning: A subset of machine learning involving neural networks with three or more layers. These neural networks attempt to simulate the behavior of the human brain to process data and create patterns for use in decision making.

Internet of Things (IoT): A system of interrelated, internet-connected objects that are able to collect and transfer data over a wireless network without human intervention.

Key Performance Indicators (KPIs): Quantifiable measures used to evaluate the success of an organization, employee, etc., in meeting objectives for performance.

Machine Learning (ML): A branch of AI that focuses on the use of data and algorithms to imitate the way humans learn, gradually improving its accuracy.

Neural Network: A series of algorithms that attempt to recognize underlying relationships in a set of data through a process that mimics the way the human brain operates.

Predictive Modeling: A process used in predictive analytics to create a statistical model of future behavior. It involves data mining and probability to forecast outcomes.

Real-Time Data: Information that is delivered immediately after collection. There is no delay in the timeliness of the information provided.

Scouting Tools: Technologies and methods used to assess the skills, abilities, and potential of players for recruitment and development purposes.

Virtual Reality (VR): A simulated experience that can be similar to or completely different from the real world, typically experienced through a VR headset.

Wearable Technology: Electronic devices that can be worn on the body, often used to track health and fitness data, and in sports, for performance and injury monitoring.

These terms form the foundation of the discussions and insights presented throughout this book. Familiarizing yourself with these concepts will enrich your understanding and appreciation of how AI is revolutionizing field hockey.

Resources for Further Study

Diving deeper into the glossary of terms opens a wide range of possibilities for advancing one's understanding of both field hockey and the role of AI within the sport. Whether you're a player, coach, or an analyst, the abundant resources available can significantly enhance your grasp of the terminologies and concepts discussed in this book.

Books and Journals: Academic literature is one of the most reliable sources for in-depth knowledge. For those looking to understand the foundational aspects of AI, books like "Artificial Intelligence: A Modern Approach" by Stuart Russell and Peter Norvig offer thorough insights. Additionally, field hockey enthusiasts can benefit from works such as "The Hockey Dynamic" by Gavin Featherstone, which discusses the modern developments in the sport.

Professional journals—including the *Journal of Sports Sciences, Sports Medicine,* and *IEEE Transactions on Neural Networks and Learning Systems*—regularly publish cutting-edge research on AI applications in sports. These journals provide peer-reviewed articles that explore everything from performance metrics to injury prevention through AI.

Online Courses and Certifications: For a more hands-on approach, online platforms like Coursera and edX offer courses on artificial intelligence, machine learning, and sports analytics. Programs such as Stanford University's Machine Learning course by Andrew Ng or the Sports Performance and Technology course by Duke University allow you to gain practical skills that can be directly applied to field hockey scenarios.

Some platform-specific courses focus exclusively on the intersections of AI and sports analytics, making them particularly useful for those looking to specialize in this niche. Certifications from recognized institutions can also add to your professional credentials, ensuring you stay ahead in a competitive field.

Websites and Blogs: There's a treasure trove of informative websites and blogs that cater to both AI and field hockey enthusiasts. Websites like *Analytics Vidhya* and *Towards Data Science* offer numerous articles, tutorials, and community discussions on AI technologies. Meanwhile, field hockey-centric websites like *FIH.ch*, the official site of the International Hockey Federation, provide valuable resources for staying updated with the latest in the sport.

Blogs run by experienced coaches and sports analysts, such as *The Coaching Blog* and *Hockey Performance Academy*, offer firsthand insights and practical advice. These platforms often feature guest posts from experts in various aspects of the game, including the use of AI tools and techniques.

Conferences and Workshops: Attending conferences can be an invaluable way to learn from industry leaders and network with peers. Events like the MIT Sloan Sports Analytics Conference and the IEEE Conference on Artificial Intelligence provide deep dives into the latest research and developments. Workshops offered at these conferences often focus on hands-on training, presenting an excellent opportunity to apply what you've learned in real-world scenarios.

Podcasts and Webinars: For those who prefer auditory learning, podcasts and webinars are excellent resources. Podcasts such as *Data Skeptic* and *The Hockey Science and Performance Podcast* delve into analytical and performance-focused topics. Webinars hosted by AI companies and sports organizations often feature expert panels discussing the latest in AI applications and field hockey innovations.

These formats can be particularly convenient for staying updated while on the go.

Software and Tools: Leveraging the right software can make a world of difference in implementing AI for field hockey. Platforms like TensorFlow and PyTorch are invaluable for anyone looking to build custom AI models. For sports analytics specifically, tools like Hudl and STATS offer robust features for video analysis and performance tracking. Exploring the documentation and tutorials of these tools can provide practical skills and enhance your technical proficiency.

Scientific Research Papers: Adopting a more scholarly approach, platforms like Google Scholar and ResearchGate offer access to numerous scientific papers on AI and sports analytics. These papers provide insights into experimental designs, methodologies, and results that can inform and inspire your own application of AI in field hockey. Moreover, the bibliographies in these papers often point to additional valuable sources.

Online Forums and Communities: Engaging with online communities can offer diverse perspectives and real-world problem-solving ideas. Platforms like Reddit's /r/MachineLearning and specialized field hockey forums offer opportunities for discussion, sharing experiences, and learning from fellow enthusiasts. These communities are often welcoming to newcomers and can be a great place to ask questions and seek advice.

Professional Organizations: Joining professional organizations can provide structured learning opportunities and access to exclusive resources. The International Society of Sports Nutrition (ISSN) and the Association for the Advancement of Artificial Intelligence (AAAI) offer memberships that include access to conferences, journals, and networking events. Being part of these organizations helps in staying updated with industry trends and best practices.

Collaborations and Mentorships: Seeking mentorship from experienced professionals in the field can offer tailored guidance and accelerated learning. Universities and sports organizations often have mentorship programs that connect you with experts who can provide personalized advice and feedback. Collaborating on projects with peers can also be a powerful way to learn and innovate.

Government and Industry Reports: Agencies like the U.S. Department of Education and private enterprises like McKinsey & Company publish reports that can provide macro-level insights into the trends and impacts of AI and sports technology. Although they might not be as specialized as academic studies, these reports often include comprehensive data and forecasts useful for strategic planning.

By utilizing these diverse resources, you can deepen your understanding and expertise in both AI and field hockey. The journey of learning is never linear, but with determination and the right tools, it can be immensely rewarding. So, whether you're just starting or looking to expand your knowledge, there's a wealth of information waiting to be explored.

Acknowledgments

Creating a comprehensive glossary of terms encapsulating nuances of artificial intelligence and its impact on field hockey was a multifaceted endeavor. It is with immense gratitude that we acknowledge the invaluable contributions and unwavering support of numerous individuals and organizations that made this section a reality.

Firstly, our heartfelt thanks go to the field hockey players and coaches who opened their training sessions and games to us, offering real-world insights and practical examples that brought theoretical concepts to life. Their willingness to share experiences and challenges provided a robust foundation for our terminology.

We are particularly indebted to the data scientists and AI specialists who took the time to explain complex algorithms in layman's terms, bridging the gap between advanced technology and practical application in sports. Their patience and expertise allowed us to present intricate ideas in an accessible manner.

Our deepest appreciation goes to the academic community, especially those who have published pioneering research in the realms of AI and sports analytics. Their scholarly articles and case studies were indispensable resources, guiding our understanding and interpretation of technical jargon.

Special thanks are due to the developers and companies creating innovative AI tools for field hockey. Their cutting-edge technologies not only revolutionize the sport but also serve as critical examples in our glossary. Their readiness to engage in dialogue and provide detailed demonstrations was crucial for our project.

Moreover, we extend our gratitude to the sports analysts and commentators whose insightful perspectives enrich the glossary entries related to game strategy and performance metrics. Their firsthand knowledge of how AI-driven insights are utilized in live scenarios was particularly enlightening.

To our team of editors and proofreaders, we owe a significant debt of thanks. Their meticulous attention to detail and unwavering dedication ensured that the glossary's content maintained both accuracy and clarity, making it a valuable reference for readers of all backgrounds.

We also acknowledge the contributions of various AI ethics experts who provided critical viewpoints on data privacy and bias. Their input was essential in framing the glossary terms that touch upon these sensitive yet crucial issues.

A special mention goes to the field hockey community at large, including fans and enthusiasts, whose passion for the game inspired us to delve deeper and ensure that our glossary was as comprehensive and engaging as possible. Their enthusiasm for understanding the interplay between technology and sports kept us motivated throughout this journey.

Finally, to our families and friends, your support and encouragement were the driving forces behind the completion of this work. Your belief in our vision and your understanding during long hours of research and writing were deeply appreciated and instrumental in bringing this glossary to fruition.

In closing, we hope that this glossary serves as a valuable tool, enhancing the understanding of AI's transformative potential in field hockey for players, coaches, analysts, and fans alike. This endeavor was a collective effort, and its success is a testament to the collaborative spirit of all who contributed. Thank you.